MW01387961

Technology Solutions: Selection, Implementation and Management

STRATEGICALLY MANAGING TECHNOLOGY INVESTMENTS TO ENSURE A POSITIVE RETURN

Kim, Congratulations on a successful merge with minimal customer impact.

Technology Solutions: Selection, Implementation and Management

STRATEGICALLY MANAGING TECHNOLOGY INVESTMENTS TO ENSURE A POSITIVE RETURN

HUGH PATTERSON

Outskirts Press, Inc.
Denver, Colorado

Technology Solutions: Selection, Implementation and Management
Strategically managing technology investments to ensure a positive return

V2.0

Outskirts Press, Inc.
http://www.outskirtspress.com

ISBN: 978-1-4327-6387-9

Library of Congress Control Number: 2010933799

PRINTED IN THE UNITED STATES OF AMERICA

This book is dedicated to Deana Black, who inspired its creation, and to Jim Stock, Tom Darovic, Jim Fecca, and Teri Patterson for their guidance and knowledge. Their help made this book a reality. Success always comes when you surround yourself with great people.

Table of Contents

Preface

This book is intended for individuals who are tasked with selecting and implementing successful technology solutions that generate a projected and measurable return. You will come away with a blueprint for how to select, implement, and support technology solutions that successfully accomplish the business goals set by your organization. While you will see a focus on technology solutions within the financial services industry, the processes are similar in other industries.

If yours is like most organizations, you have one or more existing solutions which have been implemented in the past that hinder the achievement of business goals more than they help. While it is easy to blame the vendor for making false promises during the sales cycle, the reality is that it is up to your management team to effectively research what you are buying and to fully implement it within your organization. It is also safe to say that you currently have solutions that are holding your team back and causing them to use a work-around to accomplish daily tasks. The result in either case is a loss of productivity, a loss of potential revenue, and—in some cases—compliance errors.

Using the approach suggested by this book, you will place more emphasis on selecting the most appropriate solution, implementing it thoroughly, supporting and managing it properly, minimizing the time your team spends struggling with an ineffective solution. This book will also focus on vendor relationships and the benefits of working with your vendors to be a power user of their solutions. At the end of the day, the responsibility for having the right technology solution rests on the shoulders of your organization's management team and not with your vendors. With the help of ideas discussed within this book it is time to start controlling your destiny when it comes to technology.

Disclaimer

This book is designed to provide information about the subject matter covered. It is sold with the understanding that neither the publisher nor the author are engaged in rendering legal or accounting services. If legal or other expert assistance is required, the services of a competent professional should be sought.

Every effort has been made to make this book as complete and as accurate as possible; however, there may be mistakes both typographical and in content. Therefore, this text should be used only as a general guide and not as the ultimate source on selecting, implementing, and supporting technology solutions.

The purpose of this text is to educate. The author and Outskirts Press, Inc., shall have neither liability nor responsibility to any person or entity with respect to loss or damage caused or alleged to be caused directly or indirectly by the information contained in this book.

Chapter 1

– The Process Today

If your financial organization is like most, you purchase technology solutions based on price or reputation. Vendor or solution changes are most often the result of dissatisfaction with current vendor performance, sunsetting of older solutions, or risk mitigation relating to vendor stability. In other words, change only happens when it is forced. Purchasing decisions are based on the need to replace an existing system, perceived market pressure, sales pressure by a vendor, or other external influences. Whatever the reason for your purchase, you contact vendors to let them know of your desire. You may even use an industry-standard request for proposal (RFP) to make sure that the vendor covers all of the features and functions you need. The vendors will comb through the pages of the RFP and send back an ominous document filled with information.

Once you have the list of potential vendors, you schedule the "dog and pony show" allowing candidates to come in and demonstrate their solutions. Your management team sits through hours of presentations trying to understand the difference between offerings. Following the demonstrations and question-and-answer sessions, your team gets together to carefully choose the finalists, which are then asked for references and cost proposals. You call the references that they provided, their happiest clients—who tell you that the vendors walk on water and that they couldn't be happier with their decision to go with XYZ Company. The cost proposals come in and the bidding battle begins. You pit vendor against vendor to push the costs down. Vendors offer an assortment of terms to lower the impact of the cost and throw in some "free" incentives to make their proposal look better than all the rest. Then your team uses some internal process to choose the winner.

Now that you have purchased the solution, it is time to schedule the implementation and start working toward the dreaded day of *conversion*. You follow the lead of the vendor; after all, "they do this kind of thing every day." You circle the wagons and each of the department heads manages his or her part of the conversion. The day comes and goes and all of the issues are dealt with. In time, the solution is running and the project is done. The vendor has your money and rides off into the sunset, only to be seen again when it is time to renew.

Once you start using the solution, you realize that it does not do all of the things that the vendor assured you that it would do during the evaluation process. In addition, you find out that it does not really support some of the processes that your prior system handled with great success. The result is "change"—a dirty word for most people. As each day passes, the solution becomes hated by those who were not part of the decision-making process and some of those who were. In other words, you ended up right where you started—with the same dissatisfaction with vendor and solution. How did you wind up in the same place after spending so much time and money to make the right decision?

If this scenario rings true, you are in the majority. You got the same result because you used the same decision-making methodology as countless management teams before you. The reality is that the technology purchasing decision is yet another task on an already overloaded management schedule. In many cases, the vendor selection and conversion process is so painful that it causes an organization to stay with an existing solution even though it is no longer meeting users' needs. Unfortunately, the disparity between needs and functionality can hold your organization back from accomplishing your strategic goals. The daily "work-arounds" cut into productivity. Incrementally, the cost of a poorly selected and implemented solution causes a slow degradation of efficiency.

Where did the vendor selection and implementation process go wrong? The short answer is that the decision process evolved for a specific reason: to minimize the time the evaluation committee had to spend making the vendor or solution selection. When you think about it, these large and costly technology decisions are being made in the express lane. More fundamentally, the selection process was partially delegated to the vendor candidates right after the RFP was sent out. Your organization scheduled the solution demos, but the vendors drove the discussions. You thought that you were in control, but how many times a month do you drive a solution demo and discussion? Vendors are skilled in covering the weaknesses of their solutions and things in development are sometimes in that grey area where solution concepts have not yet made it to reality.

Next comes the weeding-out process, but really the vendors who have the most polished presentations will be the final candidates. After all, they were the only ones who said that they wanted to be your "technology partners." After narrowing the list of potential vendors, it is time to call their references. Obviously, vendor-supplied reference lists are going to consist of the vendor's best customers. Think about it: When you give references for your organization, you only pick those individuals who say great things about you. Again, you put the fate of the decision-making process in the hands of the vendor. While you have more control over the pricing than most other aspects of the vendor selection process, all of the major vendors are now training their salespeople in negotiation techniques. The vendor sales team is practiced at negating customers' haggling strategies.

Once the final vendor is selected, you receive an ominous contract with all kinds of schedules and attachments. Again, the vendor is driving this part of the process, claiming that this is the standard contract and that changes are rarely made. Once the contract is signed, the most critical part of this entire process—the implementation—is frequently handed directly to the vendor. The management team, pressed for time, will minimize its involvement and in some cases simply report conversion issues as they arise.

Even well-selected solutions have little chance of success if poorly implemented. Successful implementation of any type of solution is dependent on the organization's management team. Not taking complete ownership for the management and execution of the internal implementation plan usually dooms the solution to almost certain future dissatisfaction. Most vendors are completely prepared to convert data, configure their solution, train and support your team. But what about your corporate culture, unique products and services, personalities resistant to change, and many other technical and human details unique to your organization? The vendor could not possibly account for these things. The vendor cannot and will not manage these unique changes within your organization. What they will do is implement the technical parts of the solution and have you sign off on the project.

You are now left with a poorly selected and poorly implemented solution that in time will return you to dissatisfaction with the product and the vendor. But the process could have netted different results if you had changed the way you made your vendor decision and implemented the solution.

Chapter 2
– Strategic Technology Planning

Critical Points:

- The strategic technology plan is a technology strategy that supports the organization's strategic business goals, focusing technology dollars where they will be most effective.
- Allowing any contract to automatically renew without looking for competitive pricing or increased functionality is a lost opportunity to reduce expenses or increase efficiencies.

Federal Financial Institutions Examination Council (FFIEC) examination handbooks were recently updated to include questions regarding purchases of technology solutions. These questions are designed to determine the participation of the organization's executive management and the board of directors in the technology planning and decision making process. Since most technology investments involve large amounts of capital and have the potential to negatively impact the organization's reputation and operational risk, it is critical that these decisions are based on careful planning. While the focus of this book is on the selection, implementation, and support of technology solutions and not strategic technology planning, you cannot have one without the other.

Implementing solutions that achieve the goals set by your organization begins with a well-constructed strategic technology plan. This plan is based on the goals within your organization's business plan. Even if the business plan is informal, your organization has some idea as to what it wants to achieve in the years to come. The strategic technology plan is a technology strategy that supports the organization's strategic business goals, focusing technology dollars where they will be most effective. By aligning the technology plan to the business strategy, technology management decisions are focused on what must be done to ensure that the organization achieves its business goals. Too often technology purchase decisions are based on gut feel for where technology dollars should be spent and not on a well thought out business analysis. As discussed in the previous chapter, decisions to purchase new solutions may be based on perceived market pressure, sales pressure by the vendor, or other external influences. Organizations may decide to remain with a solution to avoid the pain of change. Converting the solutions selection process into a business decision will minimize the counterproductive influences that may currently drive your technology decisions. See Appendix A for a plan outline.

At a minimum, the strategic technology plan should include the following:

- A vendor management plan, which includes an annual evaluation of current technology vendors and encompasses:
    - A process to monitor key contract dates
    - A comprehensive Gramm-Leach-Bliley Act (GLBA) risk analysis

    - A review of performance against business goals (return on investment, or ROI)

- A plan for ongoing support of existing solutions, including:

    - User support

    - Implementation and dissemination of enhancements

    - A service level evaluation

    - Required hardware or software changes

- A plan for the evaluation of future technology purchases that encompasses:

    - Business case discipline

    - Technology budgeting process

    - Compatibility and support of strategic business objectives

Vendor Management Plan – Regulators have generally been increasing their focus on the relationships that financial institutions have with their vendors. The ability of a financial institution to select and manage vendors effectively is directly related to its safety and soundness. This regulatory concern has heightened dramatically in recent years in large part because of the trend of outsourcing services and technology.

An annual evaluation of your current technology providers will force your management team to review factors critical to determining the viability of those vendors. Each vendor should be reviewed on an annual basis with the goal of lowering operational and support costs and maximizing your technology investment. Allowing any contract to automatically renew without looking for competitive pricing or increased functionality is a lost opportunity to reduce expenses or increase efficiencies. The vendor may not be able to help each time they are asked, but the exercise allows you to capitalize on opportunities as they do arise. At the very least, it puts the vendor on notice that you have an ongoing awareness of the cost and benefits of the solution. Knowing key contract dates, especially renewal notice requirements, will keep your organization from missing opportunities to review other vendor offerings and to make well thought out strategic vendor decisions. The market evaluation process must take place well in advance of the contract renewal date. Therefore, a plan must be in place to prompt the review and evaluation of costs verses benefits prior to the existing contract's termination. Your regulators will review the plan that you have in place to ensure that it appears adequate.

Section 501 (b) of the Gramm-Leach-Bliley Act (GLBA) requires that all financial institutions have a process in place to measure, monitor, and control risks to customer information. The steps outlined within this book will touch on some of the requirements as they relate to solution selection, implementation, and management. Regulatory publications dedicated to interpretation of the Gramm-Leach-Bliley Act should be consulted to understand the complete requirements of the act as it relates to your vendor management program. Not only is a comprehensive risk analysis required by your regulators, but it is also critical to determining if a vendor is healthy enough to continue as your technology partner. The analysis should include a determination concerning the criticality of the vendor (solution) to your organization's ability to operate. The vendor's strategic, operational, reputation, compliance, and financial (credit) risk should also be considered. In situations where the risks are significant, an alternate vendor should be identified as an alternative should the primary vendor weaken.

Strategic Risk – Establishing internal accountability for managing each vendor relationship is essential to forming the kind of partnership necessary to understand the future development plans of your vendor. The best way to accomplish this is to have a system in place to ensure that someone within your management team is following trends within your marketplace and the technology industry. He or she should also have a close working relationship with your vendor, allowing your team to understand the strategic direction of the vendor's research and development (R&D). Vendors should have well-documented development plans demonstrating their ability to anticipate your future technology needs. Frequently these vendors will have "road maps" of their development plans posted online or available through your relationship manager. If the vendor is not communicating this information to you, your management team should have a process in place to identify this and obtain the appropriate information. Understanding the future direction of your vendor and their solution is a critical part of the overall vendor analysis.

Operational Risk – Operational risk is an exposure to losses arising from mistakes such as computer failure, breach of regulation, fraud, and embezzlement. All of these possibilities negatively affect the organization's day-to-day business. The definition includes legal risk, which is the risk of loss resulting from failure to comply with laws, and regulations, as well as prudent ethical standards, and contractual obligations. It also includes the exposure to litigation from all aspects of an institution's activities. For this reason, each solution should be analyzed to determine what controls the vendor has in place to minimize operational risk. In addition, internal controls should be reviewed periodically to ensure that changes made by the vendor have not negatively impacted your risk mitigation strategy.

Reputation Risk – Reputation risk is the risk of negative publicity regarding your organization's business practices. Since many technology solutions on the retail and operations side directly impact your customers, this risk extends to the activities of your solution providers. For example, deficiencies in security and privacy policies that result in the release of customer information by a service provider may result in reputation damage to your organization. Your vendor management plan must have a process in place to ensure that solution providers are in compliance with the security requirements within your industry.

Compliance Risk – The organization's compliance officer should be tasked with an ongoing review of your key technology solutions to ensure that the vendor is compliant with federal, state, and local laws and regulations. Issues that are identified should be communicated to your vendor in writing with a follow-up plan to ensure that they are addressed. An unwillingness to maintain the solution within compliance standards as outlined by the contract should be grounds for immediate review of alternate solution providers.

Financial (Credit) Risk – Financial (credit) risk is determined through a review of the vendor's financial records. Audited financial statements should be obtained from critical technology vendors on at least an annual basis. The statements should be analyzed using standard credit analysis techniques and the results included in the overall evaluation of the vendor. Vendors showing a deteriorating financial condition should be monitored on a frequent basis and an alternate vendor should be identified as a contingency. The financial condition of less critical vendors can be monitored using industry publications, public market analysis, and other less intensive techniques. The degree of financial analysis should be related to how critical that vendor is to your organization's daily operations. Privately held vendors carry more risk as they do not have the same investor oversight as publicly traded vendors. For this reason, extra care should be taken during the financial analysis process to ensure that only audited financial statements are analyzed.

Review Performance against Business Goals – Solutions selected to help the organization accomplish its business goals should be measured periodically. Measuring the effectiveness of each solution will allow your management team to make adjustments to maximize the solution's performance. We will discuss this more in Chapter 13. The projected return on investment (ROI) does not always have to be quantitative, but it should be measurable in some way. Some more common return indicators are:

- Increased revenue
- Improved customer service
- Enhanced retail delivery
- Increased efficiency
- Account growth
- Account retention (market pressure)
- Regulatory compliance

Ongoing Support of Existing Solutions – Since support can easily consume 60 to 100 percent of the technology budget, each vendor's level of service (user support, enhancement development, regulatory compliance) should be evaluated periodically to ensure that they are supporting you at a level that allows the organization to accomplish your business goals. In doing so, your management team will ensure that you are minimizing operational and support costs and maximizing the effectiveness of the solution. Vendors should include service level agreements (SLAs) within their contracts. If this is not the case, your management team should negotiate with the solution provider to add an SLA, or at the very least establish internal performance criteria to be included in the contract. Without proper vendor support, your team's frustration will increase. In addition, resources within your organization may be compensating for the deteriorating vendor support. If the level of vendor support is deemed to be insufficient, the vendor should be contacted and made aware of the issue immediately. Do not wait until contract renewal to voice concerns about vendor support.

Implementation and Dissemination of Enhancements – The implementation of solution enhancements and the dissemination of enhancement information are critical to getting upgrades and system corrections out to your team. It also impacts the organization's training and technical resource projections. A formalized process to monitor and plan for the rollout of enhancements to existing solutions will ensure internal accountability for implementation. While you generally do not want to be the first to implement any type of system change, long delays in the implementation process could mean increased compliance and security risk. Failure to stay current on a vendor's hardware or software release level can also negatively impact that vendor's ability to support your team.

Your plan for implementing pending hardware or software updates should include a standard lead time unless the update is known to be critical (as in security or compliance patches). During this lead time, the internal solution sponsor should have discussions with the vendor and other vendor clients to identify any issues that may result from installation. Based on the criticality of the system and the estimated risk, you may want to assign internal team members to conduct acceptance testing. Acceptance testing will require that the hardware or software be operated in a "test environment". Ideally, the "test environment" will mirror your production system in a way that will allow you to identify any potential issues while not impacting the production environment in any way. If the solution is being operated in a data center environment, the data center should conduct acceptance testing on your behalf. Once your management team is comfortable that the update can safely be installed, the resulting changes should be communicated to users of the solution in advance of its installation. If policies or procedures are affected, these guidelines should also be updated prior to the implementation date. This will ensure that there are no surprises during the next audit.

A Service Level Evaluation – Quality user support is critical to effective use of both hardware and software solutions. This includes both internal and external support of users. Periodic internal user meetings will allow your team members to provide feedback to the management team concerning support and solution issues in general. In addition, metrics maintained by the solution sponsor (based on service level agreements or service level expectations) and the vendor should be reviewed periodically.

Required Hardware or Software Changes – As technology evolves, so do the requirements of your current solutions. Advances in hardware and programming architecture will require your vendors to alter their products to remain competitive. Required changes in hardware or software used to support a solution are rare and are usually prompted by a third-party vendor sunsetting or altering a solution, as in a teller system formatted to print on a certain type of receipt printer. Once the third-party vendor for the receipt printer decides not to support the hardware, your solution vendor has no alternative other than to certify a new printer. As a client, you would be forced to retrofit all workstations with the new hardware. When these events do occur they are usually not a surprise. Announcements by a vendor of their intent to sunset a hardware or software solution are usually made a year or more in advance.

In other situations, the choice to move forward with more advanced hardware or software is up to your organization's management team. For example, say Microsoft rolls out a new operating system. Your organization must decide if it should proceed or delay the rollout of the new operating system. If you delay the rollout, how does that impact vendor support of the current system? Is the vendor currently offering financial incentives to proceed with the rollout? Are there penalties for not migrating? Are ancillary third-party systems compliant with the new operating system? If not, when will they be? Is there an advantage to doing nothing? While this is a narrowly focused example, the organization should have a plan for the support of all technology solutions that are being used to achieve its business goals. There is little doubt that existing solutions are currently being supported within your organization. What may be missing is a formal planning process for that ongoing support.

Evaluation of Pending Technology Purchases – When a technology purchase is being considered by your organization, you should have a process in place to evaluate the effectiveness of the solution based on the achievement of key business goals. It should also include a plan to monitor the success of the solution in achieving these goals. Depending on internal triggers—such as estimated cost or how critical the solution is to your organization—a business case process may need to be initiated.

Business Case Discipline – The business case document is a justification of the expenses and operational impact related to the purchase cost and implementation expense of the new system. The discipline inherent in using a business case evaluation process is intended to ensure that your management team stays focused on solutions that support your business goals. The business case methodology focuses on the business need, anticipated outcomes, impact to the organization, risk analysis, expected return, and projected costs. We will discuss the business case document in more detail in the next chapter. For now, it is important to remember that the business case methodology should be documented within the strategic technology plan.

Technology Budgeting Process – Well-run financial organizations do a good job projecting income and expense when it comes to annual budgeting. This budgeting process occurs each year and the entire management team is familiar with the routine. The strategic technology plan is simply an added discipline within the budgeting process. Prior to beginning the annual budget process, a technology "wish list" should be created. Ideas for technology projects come from vendor conferences, trade shows, local compliance meetings, regulatory postings, knowledge of the local market, and advertisements. Each proposed technology project should be sponsored by someone on the management team. This person should have some idea as to the solution's functionality and the business need that the solution is intended to address. The management team will then review each request and discuss needs based on priority. For example, is automating the return of deposited items (a cost/risk reduction) more important than implementing electronic cash management (a revenue opportunity)? Solutions that support a valid business need can be further analyzed to determine projected cost, benefits, and operational impact to the organization. Projects that make the final cut should be considered for inclusion in the overall organizational budget. An analysis of projected cost, expenses, resource requirements, and staffing availability will allow your team to determine the feasibility of each project.

Compatibility and Support of Strategic Business Objectives – The organization should maintain an ongoing awareness of new technologies and market trends involving technology-based solutions. Do the resulting solutions affect your market? If so, can they contribute to your organization's strategic business objectives? Regulatory requirements can also positively and negatively impact current and future use of technology. The strategic technology plan should identify the individuals within your organization who have primary responsibility for this awareness as well as a plan for sharing this information with the management team.

As noted in the beginning of this chapter, the strategic technology plan is the creation of a technology strategy that supports the organization's strategic business goals, focusing technology dollars where they will be most effective. By aligning the technology plan to the organization's business strategy, technology management decisions are focused on what must be done to ensure that the organization achieves its business goals. The strategic technology plan acts as a guideline for support of current technology and the purchase of future solutions. Within the strategic technology plan there should be a vendor management plan, a plan for the ongoing support of existing technology solutions, a plan for the evaluation of future technology purchases, and a technology budgeting process. All of these plans and processes ensure that the technology expenditures made by your organization are stable, well managed, and focused on valid business goals with an achievable return.

Chapter 3

– We Need This Solution

Critical Points:

- Always ask, "What business need is this solution going to address and how are we going to recoup the cost?"
- Adding new solutions that have no measurable return simply to keep up with the competition may negate your current advantage of lower overhead.

Most decisions to buy or upgrade a technology solution begin at vendor conferences or industry trade shows. Attendees view the most innovative and comprehensive solutions ever created by mankind. Exhibitors reveal how these marvels will help attendees accomplish the previously unthinkable and they are told that anyone not using the latest product is being left behind. These disciples then return to your organization and campaign for the purchase of these solutions based on the belief that "we have got to have this solution."

While this scenario will occur, it is how the management team reacts that will make the difference between success and failure. Vendor conferences are a great way to keep up with vendor offerings. Attending them is a must if you are going to have a successful relationship with your vendors. They allow your team to stay current on developments and to see what the future holds for the solution. They are also showcases for the vendor, designed to get users excited about the latest products. Most importantly, organizations have to ask the question: Is this a solution to a valid business need or has the "need" been manufactured by the vendor? Remember, the primary purpose of any vendor's user conference is to showcase their wares. Some solutions are validated by regulatory or economic needs. Many others are created with flashy interfaces and the "need" is manufactured at the conference itself. Your management team must always ask the question, "What business need is this solution going to address and how are we going to recoup the cost?" In other words, do we really need to do this? What is the cost and what is the return on our investment (ROI)? Until the organization can answer these questions, no further action should be taken to invest in these solutions.

Understanding what business need the solution is going to address is relatively simple. Either you have an operational, financial, or a service-delivery need for which a technology solution can reduce costs, reduce errors, increase revenue, increase retention, or increase your efficiencies—or you do not. Purchases based on perceived need or political reasons (pet projects) with no clear business justification may still achieve their goal. Unfortunately, that goal may not help your bottom line. It may, however, also use internal resources that could have been deployed to positively impact your business goals. It will most certainly negatively impact the organization's efficiency as it is not focused on your business goals. In this situation, accusations are frequently made that the product was oversold by the vendor. In truth, the responsibility to properly fit the solution with a valid business need rests with your management team.

The challenge with every new technology purchase is to clearly determine and justify how it will positively impact your business strategy. In many cases a technology is adopted due to perceived market pressures and not due to something tangible, such as actual market research indicating customer demand. But why should we increase our overhead by adding additional cost to our organization without a valid justification? Another organization within our market may have implemented a solution and increased its costs with little demand or payback within the market. Their decision to purchase was based on perceived need. By using that same justification as a catalyst for making our purchase decision, we have now negated what was a true market advantage (lower operating expense) for our team. Building a well-thought-out business justification for technology expenditures reduces the decision to a black-and-white process. If the "need" to purchase a solution is based on customer demand, why not form a focus group with a sampling of your most valued customers? Ask the members for their input and validate their true needs. This will tell you if you are on the right track. If there is a valid need, it will motivate them to be champions of the solution within your market. After all, they were the ones who told you that the solution was needed.

When presented with the "need" to purchase or upgrade a solution, your management team should ask and answer the questions: Do we really need to do this? What is the cost and what is the return on our investment? In the next two chapters we will examine the cost analysis and discuss return in more detail.

Chapter 4

– How Much Is It Going to Cost?

Critical Points:

- Estimating the cost of ongoing support for the new technology solution is as important as estimating the purchase and implementation costs.
- Communications requirements should be evaluated and included in the solution cost estimate.

When it comes to cost analysis, multiple factors must be considered. Purchase decisions made based on high-level estimates of vendor pricing are frequently met by unexpected subsequent costs, potentially doubling or tripling the original project estimate. In each of the areas listed below, cost estimates should be obtained from a sampling of vendors being considered. It should be noted that at this point in the process no clear vendor will have been chosen. For this reason, all of the *estimated* costs will be high level. They will be used for the purpose of evaluating the viability of moving forward with the purchase or replacement of a solution—not for the actual contract negotiation. These cost estimates should be used for estimated project budgeting. The common areas listed below should be considered when projecting the cost of a new or replacement solution:

- Cost of solution
- Cost of implementation
- Cost of vendor support
- Conversion/deconversion of existing data
- Cost of training
- Cost of supporting software (Citrix, Terminal Services, etc.)
- Cost of supporting hardware (servers/routers/workstations)
- Communications (T1 lines, installation, monthly leases)
- Cost of interfaces to existing systems
- Marketing and promotion costs (internal and external)
- Increased internal support costs

Cost of Solution – Whether the solution is a hardware lease/purchase or software licensing, the cost should be listed as a single line item. Hardware can either be purchased with a single upfront cost and then amortized or leased with the cost spread over several periods. Software frequently comes with an annual license fee for the term of the contract period. Vendors are more likely to discount upfront software licensing fees than most other items. Hardware may also be deeply discounted to incent you to become a customer. For this reason, the cost of the software should be listed independently. Once the total cost for the solution is determined, this line item will be the starting point for future negotiations.

Cost of Implementation – The cost of the implementation includes only the cost related to the actual implementation of the solution. It will not include conversion or other third-party-related costs. Costs should be estimated for any travel related to setup, training, or project management. The solution vendor should be able to help you estimate the number of trips required and their duration. In some cases the vendor will waive the cost of implementation as part of the contract negotiation. It is best to obtain a cost estimate from the vendor when evaluating the total cost of their solution. At the very least, you will know the value of the item (even if it is waived by the vendor later) when it comes time to negotiate the contract.

Solutions that involve customer interaction (customer-facing solutions) will add additional cost to the implementation. For these types of solutions you will need to estimate the cost for educating your team related to positioning the solution with your customers, marketing, customer communication, and support. We will discuss this in more detail in a later chapter.

Cost of Vendor Support – If vendor support is not included in the upfront cost of the hardware or software, it should be included here. The vendor should be able to help you estimate this cost if it is based on metrics such as the number of support calls or other variable factors. Some software vendors offer a standard service level agreement (SLA) outlining service benchmarks. If this is an option, your management team should review the SLA to ensure that the benchmarks are adequate. Hardware vendors usually have tiered levels of support with increased cost depending on service response time and turnaround time for parts. Mission-critical solutions should warrant the highest level of service available. Do not compromise on the cost of support.

Conversion/Deconversion of Existing Data – Conversion and potential deconversion costs need to be identified and included in the cost analysis. The vendor being evaluated should provide the cost of converting existing data. If you are migrating from an existing solution, you will need to contact your current vendor for costs related to deconverting data in a format acceptable to the new vendor. In some cases, the cost of deconverting data from an existing system can exceed the cost of the new solution itself. It is best to get a formal written proposal from your current vendor for deconversion costs during the cost analysis process. At the very least it puts your current vendor on notice that you are seriously considering other solutions.

Cost of Training – The cost of training should include fees the vendor charges to provide initial conversion training as well as any ongoing training and related travel costs. Training may include both technical and user training. Travel could cover the cost of vendor personnel traveling to your location or the cost of your team members traveling to the vendor location. One alternative frequently used to minimize training cost is called "train the trainer." In this process, a few key members of your team receive initial training; these members will then be charged with training the entire organization. This is a viable alternative only if the trainers are well qualified to conduct the training and you hold the trainees accountable for what they have learned. Another training option frequently used is that of Web-based training. Web-based training is only effective when the students can extricate themselves from the work environment. They must have dedicated and uninterrupted time to complete online lessons in order for this type of training to be effective.

Cost of Supporting Hardware/Software – The infrastructure changes required to facilitate the installation and operation of the new solution may include additional servers, workstations, operating systems, communications programs, and networking software. New solutions frequently require additional infrastructure to be purchased and installed. Replacement of existing solutions may require an upgrade in hardware or software. The prospective vendor should be able to supply you with a specification document listing hardware and software requirements for running their solution in your environment. All of these costs should be determined and analyzed prior to the purchase decision of any replacement solution. Pricing can be obtained from the solution vendor as well as third-party sources to ensure that you are getting a reasonable cost estimate.

Communications – One critical item frequently left out of the evaluation process is the cost of communications. This may be the cost of data lines to your data center or other offices within your network. Existing infrastructure should be analyzed to determine if it can support the bandwidth requirements of the new solution. If upgrades are required, an estimate should be obtained from your providers. Any incremental monthly leasing or service charges should be calculated for the term of the vendor's contract and included in the total estimated cost.

Cost of Interfaces – Before getting into the pricing aspects of interfaces, we need to discuss the difference between the terms "interface" and "integration." The term "integration" means that functionality has been written into the primary solution's programming. In other words, it is a part of the system that you are purchasing. The overwhelming advantage of integrated functionality is that changes made to the primary solution by its vendor are also made to the add-on functionality. Major system releases and upgrades should be seamless and should flow through to all integrated functionality. The drawback is that it is usually an all-or-nothing situation. In other words, you cannot change the primary vendor and keep the add-on functionality.

The term "interface" indicates that the extra functionality is not part of the primary programming and that an interface will have to be used to facilitate communication between the primary system and the secondary functionality. In most cases, release and system update changes made by the primary vendor will require your third-party vendors to modify their interface coding. There is also usually a cost associated with this development.

Interfaces to existing systems may be required in order to take full advantage of the new solution's functionality. If third-party interfaces are required, your management team will have to contact the appropriate vendors to determine the cost of the interfaces. In some cases they will already exist and you will simply pay a licensing fee to use them. In other cases, they will have to be developed and certified. If you are faced with the latter, you should contact both the solution vendor and the third-party vendor to see if they are willing to pay the costs related to development and certification of the new interface. It would also be a good idea to get any agreement regarding development, certification, and cost in writing prior to signing a contract.

Marketing and Promotional Costs – Although it may seem a little premature during the evaluation phase, marketing and promotional costs should be estimated and included here. This item involves determining how large an advertising budget you need to make this solution a success in your market. In some cases a solution vendor will have marketing and promotional materials already created. If this is the case, you should determine the cost of the materials and ask for contact information relating to clients who have used the vendor's marketing plan. Any costs relating to customization should also be included in the final estimate. This will help you determine the value of using the vendor materials as opposed to developing your own.

Increased Internal Support Costs – With each new technology solution, there is a risk that operational support costs will increase. The daily operation of the hardware and software must be supported either by the vendor, internally, or through a third-party service provider. Operational processes relating to inputs (hardware usage, data input, item handling, workflow, etc.) and outputs (system reports, data files, balancing, interfaces, etc.) must also be considered. The increased costs to the organization should be estimated and considered as a cost of the solution. You may be able to offset some of the support costs if you are replacing an existing solution; however, there will be some type of support costs related to the new system.

Now that the cost estimates have been obtained, you are ready to move to the project evaluation phase. This step in the solution evaluation will use a document called the business case to match the solution benefits with strategic business needs; the result is a viable and measurable projected return on the technology investment. It will also facilitate the aggregation of the hard costs obtained in the previous steps with the soft costs related to the impact of the solution on your organization. The goal in this step is to determine if the organization should proceed with the project, and to establish measurable benchmarks that will be used later to evaluate its success.

Chapter 5

– Evaluating the Solution

Critical Points:

- The business case process allows your management team to objectively evaluate the solution costs versus the projected returns of technology projects before you proceed with the purchase.
- The business case should validate the business need being addressed by the technology solution.

Once the cost estimates have been collected, an internal team member should be assigned to create a business case for each project being considered for possible inclusion within the budget. The business case is a document that allows your management team to objectively evaluate the estimated benefits and costs of technology projects before you proceed with the purchase. The complexity and cost of the solution being considered will determine the level of analysis required to justify the decision. At a minimum, a good business case should include the following elements:

- Solution description
- Business need/justification
- Estimated impact to current operations
- Projected return on investment (ROI)
- Estimated timeline for implementation
- Hardware/software requirements
- Key resources required
- Estimated total cost
- Assessment of impact to organizational risk

Solution Description – The solution description should be detailed enough for the entire management team to understand what solution is being considered and its impact to the organization. It should also include the expected features and functionality, identify the area of the organization providing support, and pinpoint the potential users being impacted.

For example, say your organization is looking at a solution for tracking collections. The functionality could allow users to identify accounts that are ten, thirty, and sixty days past due, along with the ability to enter comments and track promises to pay. Additional functionality might include onscreen data including information on current accounts, account balances, customer names, telephone numbers, addresses, and relationships to other individuals and their contact information. The collections personnel will be responsible for viewing and entering the collections data. The account operations area will be responsible for maintaining the integrity of the account database. The IT group will be responsible for hardware and software support.

Business Need – The business need will tie the solution to the strategic business plan, indicating the business goal being addressed by the solution. It is important that the management team agree on the business need and that the proposed solution support that goal. In our collections example, the business need could be identified as increased revenue through the collection of past-due accounts.

Estimated Impact to Current Operations – The estimated impact to current operations section should discuss the projected changes to current operational processes, workflows, controls, and compliance. For example, if the implementation of the solution will involve a customer-facing product (Internet Banking, voice response, etc.), then information security, compliance, support, and operational processes regarding updates and content should be discussed here. Expected staffing increases or decreases should also be included, as well as pre- and post-implementation training.

Projected Return on Investment – The projected return on investment (ROI) does not always have to be quantitative but it must be measurable in some way. Some more common return indicators are:

- Increased revenue
- Improved customer service
- Enhanced retail delivery
- Increased efficiency
- Account growth

- Account retention (market pressure)
- Regulatory compliance

The management team must define a way in which the resulting impact can be measured along with the expected lead time for results to be measurable. In the case of our collections example, the return could be as simple as a 20 percent improvement in past-due account collection to be measured at the three-, six-, nine-, and twelve-month intervals. The return must be measurable and the goal must be reasonable.

Once the "how" is answered, the "who" must be addressed. Who in the organization will be responsible for reporting back to the committee regarding the success or failure of the project? The management team must appoint someone who will measure the return and report back to the committee as to the progress toward meeting the return goal.

Estimated Timeline for Implementation – In order for the organization to properly budget both time and resources, the estimated time/resources portion of the business case should be as accurate as possible given the early stage of the process. A telephone call to prospective vendors will likely net some level of detail necessary to estimate these costs. Most vendors will have a recommended implementation timeline depending on the scope of the project. The timeline should include time budgeted for the following:

- Preproject planning
- System configuration – planning/review
- Training of users
- Installation of hardware/software

- Testing of conversion files
- Conversion of existing data/review
- Interface testing
- Solutions rollout/promotion
- Marketing
- Project wrap-up

Only your management team will be able to estimate internal resources required to implement or convert a solution. See Appendix B for a business case worksheet.

Hardware/ Software Requirements –This section of the business case will detail the actual hardware and software required to operate the solution that will be purchased. It will also include an analysis of how that might impact existing infrastructure. Although the business case is intended to be an evaluation tool and not a detailed project plan, it is important to give some thought as to how the proposed hardware and software changes will affect your current configuration. What is the net impact of what you are proposing compared to your current solution? For example, what would you do with existing hardware and software if you elected to outsource a replacement solution? Can you sell it or use it for another solution? If the equipment is being leased, can you break the lease? The answers to these questions will give your management team a more complete analysis of the hardware and software costs associated with the pending project. The vendor's specifications documents can be used to determine what infrastructure changes are required.

Key Resources Required – This section estimates the involvement of key individuals within your organization critical to ensuring that the solution is properly implemented. Doing so will ensure that internal resources are aware of the commitment expected of them should the project be approved. It will also help to prevent your organization's key individuals from being overextended should there already be planned commitments for their time.

Estimated Total Cost – The estimated total cost section will incorporate the detailed cost information gathered in the previous chapter into the business case document.

Since no vendor has been selected at this point, you will not have a complete total cost figure. This estimate, along with the soft cost estimates within other areas of the business case, will allow your management team to begin the evaluation of the solution's estimated benefits versus its estimated costs.

Assessment of Impact to Organizational Risk – Projected impacts to strategic, operational, reputation, compliance, and financial risk should be estimated and summarized in this section of the business case. This does not have to be an in-depth analysis, but it should be thorough enough to allow the management team to properly evaluate the risks associated with the project. For example, the implementation of a social media marketing plan carries with it a high degree of reputation risk. Poorly handled customer service issues that were once contained to word of mouth will now be broadcast to the world.

A deadline for completed business cases must be set to ensure that the evaluation process is completed prior to the budget process. Frequently the business case process itself will cause an organization to abandon a project. Either the sponsor deems the business case process too burdensome, there is a realization that the cost of the solution itself is too high, or the hardware/software requirements appear unreasonable. Whatever the scenario, the sponsor should be encouraged to complete the business case even when he or she would tend to rule out the project. Subsequent committee evaluation may validate unexpected costs or requirements when they are analyzed against the estimated benefits. Once the business case documents are completed by their sponsor, they should be forwarded to the management committee members for review. Committee members should review these documents as soon as possible to give feedback to the sponsor on areas that may need additional research. Updated documents should be forwarded to each committee member as soon as they are complete.

The business case review meeting should be an open forum where each committee member has an opportunity to voice concerns and objections as well as desires for expected outcomes. Whereas one area of the organization might benefit greatly from the implementation of a particular solution, another may see additional costs in the form of increased operational or support requirements. The underlying questions are: What is the return on this solution? How will we measure the return? And are the costs outweighed by the benefits in terms of their ability to help you achieve your business goals? Projects that pass this test should be considered a viable candidate for inclusion in the budgeting process. Your management team knows its budget for capital expenditures and human resources. Only projects that fit into these resource parameters should be approved for inclusion in the coming year's budget.

It should be noted that when it comes to the evaluation of replacement systems, the cost of doing nothing should be carefully evaluated. If the existing solution does support your organization's strategic business goals and you are simply doing "due diligence" as per your technology plan or vendor management program, the cost of doing nothing can be low. If, on the other hand, your current vendor is not supporting your strategic business goals, the cost of doing nothing can be high. Daily "work-arounds", user frustration, and the inability to offer competitive products to your customers carry incremental costs that add up to lost revenue and decreased productivity.

When looking at new technology solutions you must look at the opportunity cost. Depending on how new the technology solution is, vendors may be looking to make special deals to get users in the market. If the solution life cycle is past the introduction phase, you may be looking at higher vendor pricing as demand grows. Obviously, the higher the pricing, the more important it is to look for the business benefits or return necessary to offset the cost. For those risk-averse individuals within your organization who would rather hold off on the purchase of new technology, (let's wait and see) the cost can be lost revenue, lost market share opportunities, and a reputation for always being behind. Only you can determine if the benefits outweigh the costs.

Chapter 6

– Should We Outsource?

Critical Points:

- The option to outsource should always be included in the solution evaluation process.
- Corporate culture will significantly influence the outsourcing decision.

Consideration should always be given to using a "service provider" to outsource the solution under consideration. In some cases, the primary vendor will own a data center where their solution is operated and supported. In other cases, third-party service providers operate independent data centers running the solution. Even if your corporate culture tends to prefer running solutions in-house, there are advantages to considering outsourcing as a way of decreasing operating and support costs. In addition to corporate culture, some of the variables that can impact the outsourcing decision are service provider availability, logistics, cost savings, and regulatory compliance.

Corporate Culture – Organizations tend to favor either running all technology solutions in-house or outsourcing all solutions. Typically, smaller organizations operate in an outsourced environment while larger organizations have the staffing and desire the control that comes with running solutions in-house.

Service Provider Availability – Large time zone differences between your organization and the service provider should be avoided. It may not seem like a significant factor when the vendor-added financial incentives are included in the decision model, but large time zone differences can negatively impact your service provider's availability and their ability to support the solution. Even when you are assured that the after-hours support team can handle any crisis it is best to make sure that the vendor references concur. Critical issues discovered at the end of the workday can go unresolved until the next business day when there is a large mismatch between the vendor's and user's primary operating hours.

Cost Savings – Cost savings is the reason given most often for outsourcing a solution. Service providers use economies of scale to decrease their operating costs while providing your organization with greater operational bandwidth and increased solution expertise. While outsourcing is a cost-effective way to operate a technology solution, transaction, processing and handling fees can easily negate the savings and actually increase the overall cost.

Regulatory Compliance – With increasing regulatory oversight concerning customer information security and a host of ever-changing regulations, service providers must maintain a constant state of awareness and compliance. A data center serves multiple clients; each customer reviews the center's controls and compliance practices on a regular basis. In addition, regular external audits and annual Statement on Auditing Standards No. 70 (SAS 70) reviews ensure that the service provider is operating within the guidelines set by the laws and regulations. Of course in the eyes of your regulators, the use of a service provider does not absolve your management team of the responsibility for compliance. But it does defer the job of regulatory interpretation and implementation to a third party. It is important to note here that you should always make sure that service providers are aware of and in compliance with the laws and regulations unique to your state.

While the advantages and disadvantages can vary by solution, the following points should be kept in mind when consideration is given to each option.

Advantages of Operating the Solution In-House

- Control – Operating a solution in-house gives you complete control over its operation, security, the input, and the output. You also control when solution updates and fixes are installed.

- Staffing – Operations and support staff are hired by and dedicated to your organization. They know your corporate culture and have a vested interest in the organization's success.

- Depreciation – Hardware and software purchases are depreciated rather than expensed.

Disadvantages of Operating the Solution In-House

- Overhead – Operating a solution in-house requires that you purchase the proper hardware and software, and hire the human resources necessary to operate and support the solution.

- Security – Your organization is completely responsible for physical and system security.

- Business recovery – You must develop and manage your own business recovery plan.

- Compliance – All responsibilities for compliance and risk analysis fall on the shoulders of your team.

Advantages of Outsourcing the Solution

- Expertise – Service providers specialize in their field. These vendors also have specific hardware and software expertise. The tasks can generally be completed faster and with better quality output.

- Overhead – Outsourcing the solution minimizes the need to add hardware, software, and human resources in order to support a new solution.

- Greater resources – Service providers have greater depth in operating and supporting the solution. Their equipment can also handle growing and peak system volume.

- Risk sharing – Risk is shared with the outsourcing vendor.

- Business recovery – The service provider should have certified business recovery plans in place.

Disadvantages of Outsourcing the Solution

- Risk and security controls – You have less control over risk and security controls.

- Quality control – You have less direct control over quality. Your management team controls input but not output, operations, and support.
- Service provider stability – The possibility of the service provider being acquired or going out of business is out of your control.
- Lack of independent focus – An outsourced vendor is dedicated to the needs of multiple organizations. This means that they are unable to focus solely on the needs of your organization.
- Cost – Depending on competition, market conditions, and operating efficiencies, outsourcing may be more expensive than an in-house solution. Frequently, handling fees are added to all nonstandard tasks.

Even if your organization has a history of running all solutions in-house or outsourcing all solutions, pricing for both options should be requested in the request for proposal (RFP). Doing so will allow your organization to make a proper evaluation of the costs associated with operating the solution in both environments.

Chapter 7

– The Request for Proposal (RFP)

Critical Points:

- The request for proposal (RFP) is a document created by the organization to facilitate the evaluation of vendors and their solutions.
- Well-thought-out and detailed RFP documents are critical to saving time later in the evaluation process.

Once a business case is approved and included in the budget, the organization moves into the vendor evaluation process. The prospective vendor list can be compiled several different ways. If you have access to a consultant, he or she should be able to provide you with a list of potential vendors. Other ways to compile a list is to visit industry trade shows or to call other organizations within your industry and ask them whom they are using. If you are unsure whether your potential vendor list is appropriate, the business case sponsor could produce a request for information (RFI). An RFI is simply a brief description of the solution being investigated along with a request for the potential vendor to indicate an interest in participating in the evaluation process. The RFI would be sent to potential vendors with a deadline for the response. The important rule to remember here is to keep the RFI brief. Vendors are unlikely to respond to lengthy RFI documents.

A request for proposal (RFP) is a document created by the organization to facilitate the actual evaluation of vendors and their solutions. The main purpose of the RFP is to communicate the specific requirements of your solution, needs, and objectives to prospective vendors in an organized manner and to communicate expectations of the vendor selection process. Generally, the more effort put into creating a thorough RFP, the less time your management team will have to spend sifting through responses for solutions that do not address your organization's needs. At a minimum, an RFP should address the following:

- Statement of purpose
- Background information
- Operational requirements
- Technical requirements
- Support requirements (hours/time zones)
- Performance standards
- Security requirements
- Interface requirements
- Business recovery options
- Required deliverables
- Special contractual requirements

- Pricing structure
- Vendor qualifications and references
- Financial analysis
- Administrative expectations
- Organizational contact information

Statement of Purpose – The statement of purpose should be based on the business case document. It should state the objectives of the solution and what business goals are to be accomplished by the purchase. The more specific your stated goals, the better potential vendors will understand what your organization is looking for. This will give vendors the chance to match features with benefits, making the evaluation process less difficult for you. For example, your organization is looking for an Internet Banking solution that incorporates bill payment and account aggregation with the goal of increasing revenue by 5 percent and increasing customer retention by 20 percent.

Background Information – Background information will include anything that would help a prospective vendor understand your organization and your culture. A discussion of your organizational history, mission statement and market demographics would be included here. This information will also help the vendor to focus its solutions on your specific needs.

Operational Requirements – Details concerning how the solution should operate will be included in the operational requirements. These requirements, also known as "functional specifications," will include such things as workflow, transaction types, transaction volumes, operational handling, user interfaces, balancing, and regulatory requirements. Plenty of time should be allotted for managers to define the operational requirements of their respective areas of responsibility. The testing of system updates and releases should also be addressed here. If your organization would prefer to test all releases in a "test environment" prior to putting them into the production environment (acceptance testing) this requirement should be included here.

Technical Requirements – The technical requirements will be determined by the type of hardware, network, communications infrastructure, and operating system that your organization currently uses. It is crucial that the potential vendor understands your current infrastructure and any plans to change this configuration. The vendor's system may or may not fit within your current framework. If it does not, this will allow all parties to understand what hardware and software changes will be required.

Support Requirements – The support requirements section will include any specific support needs that your organization has. This would include business support hours, 24/7 user support, holiday coverage, after-hours call center support, same-day parts service, etc. Be aware of vendors' service locations and the impact that this may have on their ability to support your organization. Hardware support should be based on how critical the solution is to your ability to operate. The more critical the hardware, the more quickly technicians and parts need to be available.

Performance Standards – Any performance expectations that your organization has should be included in the performance standards section. This could include 98 percent uptime or same-day processing of payments or electronic transactions. This information will come from the functional areas of the management team. This section is particularly important when the solution being considered has the potential to increase workflow and ultimately the need for increased solution bandwidth. In these cases, the benchmark for the performance expectations should be based on the projected volume and not the current volume. The performance standards identified here should be included in the final contract.

Security Requirements – Expectations relating to security concerns—including physical, online, data storage, encryption, transmission, and other customer information security issues—should be included under the security requirements section. This would encompass security standards required by policy or regulation. The security requirements identified here should be included in the final contract.

Interface Requirements – The section on interface requirements will include a list of vendors currently being used by your organization that might in any way interface with the solution under consideration. An interface includes any third-party hardware or software now operating within your organization that will provide input to or use output from the solution being addressed by this RFP. Interfaces typically involve the transfer of data from one system to another via any one of a number of layouts or interface protocols. Interface specifications should be obtained from all third-party vendors identified above. This information should then be forwarded to the prospective vendor for confirmation that an appropriate interface exists. This confirmation of an existing interface *must* be documented in writing.

Business Recovery – Regulators require that you have a tested business recovery process in place for mission-critical and mission-necessary solutions. If you are operating the solution in-house, some solution vendors offer their own recovery solution or they have an arrangement with a third party to offer business recovery services.

The solution vendor must also have a documented and tested business recovery plan for its operations. If you elect to outsource the operations of this solution to a service provider, the outsourcing vendor must have a recovery process in place. The process should be documented and tested at least annually. The service provider should also include your organization in the recovery testing process. Unless the recovery exercise includes an "end-to-end" test of the connectivity from the service provider to your user workstation, you have not conducted a complete recovery test.

Required Deliverables – The required deliverables section of the RFP will list the items that you expect the vendor to deliver. This also includes the timeline for the project, if there is one. An example for a voice response system (VRU) would be as follows:

- Delivery and installation of VRU hardware
- Mapping of current or proposed customer prompts
- Voice recordings of customer prompts and corrections, if required
- Testing of voice recordings according to mapping
- Conversion of existing customer PIN data
- Pre-live testing using existing customer data
- Post-implementation support for ninety days
- Twenty-four-hour hardware and software support for term of contract
- Training of users

Special Contractual Requirements – The special contractual requirements section of the RFP will include any special contractual conditions. This would include such things as a contract term limitation, any special financing arrangement, or contingent acceptance of any new vendor contract. For example, vendor contract acceptance is contingent upon being able to terminate an existing vendor contract. Placing this information within the RFP alerts potential vendors of special circumstances that will need to be considered if they win your business.

Pricing Structure – The pricing structure section of the RFP should break down the solution pricing into components that will allow the management team to adequately compare apples to apples. At a minimum, the pricing breakdown should include the following:

- Software licensing
- Hardware (lease/purchase)
- Required operating system
- Required communications/data transmission infrastructure
- Installation/implementation of solution
- Training
- Project management
- Post-conversion support
- Ongoing solution support
- Required third-party interfaces

- Ancillary vendor add-ons
- Outsourcing option
- Annual maintenance or licensing fees

Vendor Qualifications and References – Information relating to current and past vendor and solution performance should be requested within the vendor qualification and references section. This would include:

- A brief history of the vendor (if not an existing vendor)
- A description of the vendor's corporate culture
- Vendor's mission statement
- Vendor's strategic development goals (development road map)
- List of references (complete customer list if vendor will provide it)
- Current financial statements
- Information concerning current or pending litigation
- Number of clients currently using the solution
- Age of the solution

Financial Analysis – The financial analysis portion of the RFP should include a current and audited copy of the vendor's income statement and balance sheet. These documents should be analyzed using accepted credit standards such as peer, trend, and cash-flow analysis.

Other indicators of the vendor's financial strength involve industry publications and discussion with existing vendor references.

Administrative Expectations – The administrative expectations section will outline the expectations that your organization has relating to the vendor's response to your RFP. These requirements should include:

- Where and to whom the response should be sent
- Timeline for submission of response
- Any format requirements
- Next steps in the evaluation process

Organizational Contact Information – The organizational contact information portion of the RFP is necessary to ensuring that the vendor is able to adequately evaluate and respond to your request. Frequently, only one individual within the organization is given as a main contact. In smaller organizations this may be appropriate. In larger organizations, a primary contact should still be listed; however, it should also include the individuals knowledgeable of the current infrastructure and operational requirements. For example, if you are looking at a human resources software solution, the managers responsible for human resources, information technology, and accounting will all need to be included. This will ensure that the vendor's technical team will have access to the individuals within your organization who can answer specific questions. These questions could concern how your operations are currently set up or future expectations for the solution.

Once the vendor responses to the RFP documents have been received, it is time for your management team to analyze the results. Each response should be reviewed to make sure that the solution being recommended by the vendor matches your organization's business case. If it does, the vendor may be considered a finalist. The review group should never deviate from the business case to make a vendor solution a finalist. If a solution causes your management team to question the business case, the RFP review should be halted and the business case itself re-examined. If the fundamental framework of your business case is sound, the process should focus on how the solution is going to support your strategic business goal. The vendor either does or does not qualify to be a finalist based on this criterion. Any unanswered questions raised during the RFP discussions can be addressed by contacting the vendor immediately. The RFP review process should be kept as brief as possible and it should not be prolonged due to unanswered questions. If an answer cannot be obtained quickly, the vendor should be rejected and the process should move on.

After compiling a list of finalists, your team will contact these candidates to schedule presentations. Often called the "dog and pony show," this is the process where vendors showcase their solutions, usually on-site. Each demo should be no longer than three to five hours; all of them should occur over the course of one or two days. There should be enough time between presentations for your team to regroup and discuss the pros and cons, but not so much time that you lose your recall of the presentations in aggregate.

Preparation for the demo process is one of the most critical parts of vendor selection. A list of questions should be created by the management team based on the business case, the vendor's response to the RFP, operational structure, corporate culture, and unique market requirements. Each department head should also view the new solution in terms of the impact that it will have on his or her area of responsibility. Some questions to consider are:

- Does this solution support our business goals? How?
- Does this solution fit within our operational structure?
- What procedural changes will be required by this solution?
- What corporate changes will be necessary if any?
- Who will support the solution and how?
- What existing solutions must interface with this solution?
- What existing third-party solutions could be replaced by integrated solutions?
- What security controls are going to have to be put into place?
- Where are the checks and balances?
- What specific benefits will this solution provide?
- What unforeseen problems or expenses exist?
- What additional risk is associated with this vendor/solution?

During the demo process, statements of cutting-edge functionality should be met with questions related to how this functionality works. What interfaces and ancillary systems are required to accomplish the functionality that you are seeing? Is this functionality available to users today? Are you viewing a production version of the solution or is it a "demo" version?

Chapter 8

– Vendor Selection

Critical Points:

- A logical evaluation based on how well each solution meets the organization's business need is imperative to purchasing a solution that will produce the expected return.
- Decisions made purely on the basis of cost overlook the potential increased revenue, productivity improvements, or market penetration gains that may be realized if the decision criteria were more focused on relevance.

Once the demo process is complete, it is time to compare notes. The management team should reconvene to discuss members' evaluations of each vendor. This can be done using a spreadsheet that gives each vendor a score based on the items listed in the previous chapter or a similar analytical process. It should not be based solely on a "gut feel" for the solution that would best fit within the organization, the flashiest demo, prettiest screens, or the "coolest" features. A logical evaluation based on how well each solution meets the organization's business need is imperative to purchasing a solution that will produce the expected return. Some additional factors that are relevant in vendor selection include:

- Solution relevance
- Reputation
- Terms and pricing
- Support
- Financial stability
- Compliance
- Security

Solution Relevance – Your management team's primary concern in the selection process should be relevance. Does the vendor offer solutions that support your organization's business strategy? Decisions made purely on the basis of cost overlook the potential increased revenue, productivity improvements, or market penetration gains that may be realized if the decision criteria were more focused on relevance.

Reputation – A vendor's reputation is a key indicator of their ability to be a good technology partner. The vendor must be willing to work with the client to ensure that the solution supports organizational needs, user issues are being addressed, and solution development plans parallel future client requirements. Does the vendor look at their relationship with you solely as a source of revenue? Or are they genuinely investing development dollars in solutions that will support your business strategy? Selecting a vendor with a marginal reputation could doom your management and team members to years of grief. Does the potential vendor have a good reputation within the industry and are their current clients happy with the company's past performance? No one knows the vendor quite like their clients. It will therefore be necessary to contact the vendor references to understand what role the vendor has chosen to play in the success of its customers. If serious consideration is being given to outsourcing the solution, vendor references will be key to ensuring that the vendor provides quality service to their clients. In contacting vendor references you should have three strategic goals in mind:

- To determine how happy they are with the solution
- To determine how happy they are with the vendor
- To obtain contact information for other clients using the product

The first two are probably obvious but the third may not be. By asking for additional user contacts, you are looking for more objective input relating to the vendor and their solution. After all, the vendor is not going to include anyone other than their best clients on the reference list. By asking for contact information from other organizations using the solution, you are going to get a more balanced view of the vendor relationship. One word of caution here: It is very possible to stumble upon a client that has a jaded view of the vendor. For that reason, it is best to try to understand why the client feels this way. Then try to contact at least two additional clients not on the reference list to determine if they share any negative experiences. Some of the questions you may want to ask vendor clients are as follows (for a complete list of questions refer to Appendix D):

- How long have you been a customer?
- How would you describe your relationship with the vendor?

- Is the vendor responsive to your support needs?
- How long do support calls typically go unresolved?
- What has your experience been with after-hours support?
- How well did the vendor work with your team to implement its solution?
- Does the vendor sponsor training? If so, is it adequate?
- Does the vendor proactively send out information on features and functionality?
- Is the vendor responsive to requests for enhancements?
- How has the vendor contributed to your success?
- Has the vendor been responsive to changes in regulations?
- If you could do it over again, would you still select this vendor?

Answers to these questions should be well documented so that others may review them.

Terms and Pricing – The terms and pricing are obviously important when you are attempting to minimize your technology costs. In some cases, your organization may be willing to sign a longer-term contract if it means lower annual cost. While it is rarely a good idea to sign a technology contract extending the relationship beyond three years, in some cases this may be necessary in order to match the cost of the solution to your budget. Vendors know this and their best terms come with five- and seven-year contracts. The risk associated with a long-term technology contract is unique to each solution. Generally speaking, the newer the technology, the higher the risk in committing your organization to the same solution for what is an eternity in the technology world. It is usually best to consider the shortest-term contract that will work within your budget.

Support – In some cases support can be weighted just as heavily as the factors mentioned above. If your organization is located on the East Coast and you are looking at a vendor headquartered on the West Coast, you must ensure that they have user support available during your normal work hours. If the potential solution is a 24/7 product (Internet Banking, bill pay, voice response), the vendor needs to provide support to you and your customers during this time period. Existing vendor clients are a great source of information concerning vendor support, especially after-hours support. They know better than anyone if the vendor is weak in this area.

Financial Stability – Another factor in vendor selection is a review of the company's financial stability. Your regulators will expect that you go through this exercise annually for all critical vendors. Because of this, you should start the selection process by choosing only financially stable vendors. A potential vendor's financial status is usually determined by doing a high-level credit analysis. A good rule of thumb is to look at the criticality of the solution. The more critical the solution is to your organization's daily operations, the more important it is for you to take an analytical look at the vendor's financial statements. With privately held companies, you may have to work with the vendor to satisfy any financial concerns that you may have.

Compliance – Most technology vendors in the finance industry are proactive about their compliance with existing laws and regulations.

If you are aware of unique state or local requirements, you should include this in the compliance analysis. Does the vendor make any written guarantees that their solution will be kept compliant with all federal, state, and local laws and regulations? If the vendor will not take ownership of this task, someone within your organization must. Vendors not willing to commit to supporting local laws and regulations should reduce the cost of their solutions to offset the cost of internal resources required to ensure compliance.

Security – The security of customer information has remained a focus of regulatory attention for the past decade. Section 501(b) of the Gramm-Leach-Bliley Act (GLBA) requires each financial institution to protect the privacy of its customers and to protect the security and confidentiality of nonpublic personal information. Because of the risk associated with breaches of security, it is critical that your management team examine the security controls in place at any potential vendor. In addition to reviewing information provided by the vendor, a Statement on Auditing Standards No. 70 (SAS 70) should also be requested. The SAS 70 report indicates that the vendor's security controls have been audited according to standards set by the American Institute of Certified Public Accountants. The report is generally divided into two types of reports: Type I and Type II. A Type I report examines the service organization's security controls in place at a specific point in time (as of the audit date). A Type II report not only includes the service organization's security controls, but also detailed testing of the service organization's controls over a minimum six-month period. The SAS 70 should be obtained and reviewed by your management team to establish that the vendor's security controls are in line with your organization's risk tolerance.

Once your management team has evaluated and scored all relative factors, a small number of vendors will likely be contenders for the contract negotiation phase of the selection process. While you may have a clear preference, it is best to engage at least two viable vendors in the contract negotiation process. This will keep your preferred vendor motivated to give you its best price.

Chapter 9

– Contract Negotiations

Critical Points:

- Vendor negotiations should focus on obtaining the maximum relevant functionality for lowest total solution cost.
- Team members who let the vendor know that they are the chosen provider prior to the negotiation process negates your advantage as a purchaser.

Before we begin this chapter, it should be noted that the information presented in this book will cover aspects specific to the negotiation of technology contracts. We will not discuss the negotiation process or negotiation techniques. If your organization is weak in this area, consult other publications that focus on the negotiation process or enlist the help of a consultant who specializes in contract negotiations.

Unfortunately, too many technology negotiations are focused solely on price and not on the larger picture. Vendors have a variety of ancillary products and services that can add value without increasing the total cost of the solution. These are things that they may be likely to concede to protect their pricing. The total solution cost should be the focus—not solely the pricing of the hardware or software. Some of the ancillary services that should be considered are:

- Implementation services
- Training services (on-site versus vendor classroom)
- Development to support existing or desired functionality
- Vendor resources to optimize usage of solution
- Third-party interface development
- Unassigned professional service hours
- Data conversion
- Additional functionality

Whether you have one clear winner from the evaluation or several, always have at least two potential vendors targeted for the contract negotiation process. Organizations that start out the negotiation with the intent of beating up on the vendor will not be winners in the end. While this may sound like a beneficial scenario, it will only start the relationship off on the wrong foot. The negotiation process should end with both parties benefitting. You pay a reasonable price for the solution, including all ancillary services, and the vendor sells their product at a reasonable price. Of course, neither of you starts out here. Even if you are already good at the negotiation process, there are some things that you need to know; we will call them "secrets of the industry":

- Only major vendors such as Microsoft and Apple set a price and demand it. The rest are too concerned that you will select someone else. Incumbent vendors are especially vulnerable when a second vendor is involved.

- Technology vendors put immense pressure on their employees to retain existing clients and to gain new business. Concessions are commonplace. Existing vendors will do almost anything not to lose a client.

The following things work against you in the negotiation process:

- Team members who let the vendor know that they are the chosen solution provider prior to negotiations do great damage. Once a vendor is aware that they have been chosen, there is little motivation for concessions.

- Unlike a car or a house, software has no defined price structure. Once the cost of the initial development is covered, vendors simply need to cover additional research and development along with support. Pricing for software is somewhat subjective with no defined ceiling or floor.

The good news is that the last two items can be controlled with a little planning. Make sure that your team members know that they should not talk to the vendor about the decisions of the selection committee. This is a major concern with incumbent vendors as they may exploit personal relationships to obtain information.

The second item is a lot more difficult to overcome. If you are working with a consultant, he or she should have some idea as to what is normal pricing for the solution. Another option is to ask existing clients what they are paying. More than likely, they are contractually prohibited from discussing pricing by the vendor but it does not hurt to ask the question. If you are given any information related to pricing, be aware that you should not disclose the source to anyone. If talking to references nets little in the way of results, the next course of action is to look at the pricing provided by the vendor. The vendor will have priced the solution higher than their target, so your mission is to determine how high. That way you can calculate the vendor's bottom line.

When negotiating a technology contract make sure that you have the following items covered as they relate to cost, management, and resources:

- Primary software licensing fees
- Primary hardware costs
- Operating system software licensing
- Networking and communication costs (if applicable)
- Implementation of primary software/hardware
- Conversion of existing data (if applicable)
- Third-party interfaces (if applicable)
- Training by the vendor
- Travel for implementation and training
- Implementation project management
- Testing/test regions/acceptance testing
- Annual maintenance/support or related expenses
- Regulatory compliance

One important item frequently left out of the technology contract is that of post-implementation follow-up. With newly purchased solutions, the organization does not always initially know what questions to ask. In addition, processes and procedures may need to be adjusted in order to use the hardware or software as it was designed. Only after using the solution for a period after the conversion will these questions and changes surface. The vendor should have resources who can conduct an on-site usage review to identify procedural and process changes as well as training that will optimize your organization's usage of the solution. There may be travel costs associated with this type of project but the costs are outweighed by the benefits of having a properly implemented solution.

The cost of third-party interfaces that are not part of the primary solution contract cause unexpected surprises. Anytime you are looking at a solution that is of critical importance to your operation, you should contact all third-party vendors to see if there are unforeseen impacts to your operations. If possible, get all responses in writing. If the solution implementation will require interface development or setup changes, you should contact both the solution vendor and the third-party vendor to see if they are willing to pay the costs related to these increased expenses. If the implementation will require a new interface to be written, it may be in the best interest of either vendor to fund the development of the new software so that they can resell the interface going forward. Get any agreement in writing prior to approving the contract.

Have all agreements reviewed by legal counsel before signing regardless of the relationship with the vendor.

Chapter 10
– The Implementation Process

Critical Points:

- Your management team—not the vendor—is responsible for successfully implementing the solution within your organization.
- The individual who will eventually support the solution internally should be intimately involved in the user training process.

Now that you have negotiated a successful contract with the selected vendor, it is time to focus on the implementation process. The single most important thing to remember here is that your management team—not the vendor—is responsible for successfully implementing the solution within your organization. It is vital that your management team assign an internal project manager to the implementation. This person will manage tasks critical to the project's success, oversee the deployment of internal resources, and provide project status reports to management. The vendor will more than likely have their own project manager, but his or her responsibility does not extend to the unique tasks within your organization unless you have contracted for this specifically. Unless otherwise contracted, your organization owns the following aspects of the project:

- Approval of the implementation plan created by the vendor
- Making database changes as suggested by the vendor
- Training of your support personnel and your users
- Coordination of internal resources necessary to the implementation
- Testing of conversion results and making/communicating needed corrections
- Ensuring that other vendor interfaces are completed and tested
- Halting the implementation if things are not going as planned

Approval of Plan – Obviously the vendor will have a knowledge base of prior implementations. For this reason, they should be expected to clearly communicate all necessary implementation activities to your management team. This is most often part of a project plan indicating the scope of work (SOW) that is being undertaken along with accountability for these tasks. The vendor should submit the project plan and SOW to your management team for approval. This will typically include the date that the project will begin, the milestones that must be accomplished to keep the project on schedule, a list of those responsible for each task, and the actual conversion date. The SOW will also include the "deliverables," or the items that you can expect the vendor to accomplish during the implementation. Your management team should carefully review this plan to make sure that you agree to the timeline and responsibilities assigned to your team. Keep in mind vacations, holidays and other community or local events that may have already been planned.

Database Changes – There is a technology saying: "garbage in, garbage out." This is most certainly the case with implementations. If your existing database contains errors such as duplicate records, inaccurate entries, or stale information, these errors will simply be passed to the new solution by the vendor. Your management team must commit to reviewing and correcting all data records prior to the conversion date. This must be done prior to or at least very early in the implementation process; doing so will save your team from future headaches. The vendor will handle conversion of the data to a format compliant with the new solution. Your management team should review the vendor's suggestions concerning data format or content. If changes are required, accountability should be established and included in the project plan.

Training – The responsibility for training your organization lies with your management team. This statement might seem like common sense, but the most common reason for user dissatisfaction with a solution after the implementation is poor training. A plan should be developed early in the implementation process indicating how and when training should occur. The individual who will eventually support the solution internally should be intimately involved in the user training process. This will ensure that when he or she hears the inevitable comment, "we were never trained on that," the trainer will be able to validate or dismiss the comment. By having this person involved with the training from the onset, he or she will begin to build knowledge of the solution and will be equipped to supplement the initial vendor training.

Training on the new solution should be addressed in the purchase contract. The vendor may require that training be done at its facility or at your location, using classroom or Internet courses. When considering a training venue, your management team must keep in mind the subject matter and how critical it is for the trainee to be focused. Depending on the complexity of the information being taught, it may be better to take staff members out of their local environment and send them to a location dedicated to training. If your organization has a training facility that is isolated from the work environment, on-site training may be the best choice. The timing of the training is also critical. Vendor-facilitated training followed by repeated practice is the optimal scenario. If this is not possible, you should plan supplemental follow-up training sessions immediately prior to the conversion. It is also a good practice to overestimate the amount of training time and resources required for the project. Rarely do implementation projects fail due to overtraining, but many fall short due to poorly implemented or rushed training.

Coordination of Internal Resources – Your organization's project manager should coordinate and facilitate regular internal meetings with primary staff members to ensure that information, accountabilities, and timelines are communicated to all levels of the organization. Feedback should be solicited so that any staff objections can be addressed before they negatively impact the implementation. One technique frequently used is to create a spreadsheet based on the original project plan. Each task is assigned to a team member and the list is discussed during each meeting.

Testing and Corrections – Implementations involving existing data will require that the current data fields be mapped to the new solution. In some cases the vendor can accomplish this, but in most cases an individual within your organization who is familiar with the application being converted should be involved. On large-scale implementations, a test bank may be used to facilitate this task. On smaller projects, spreadsheets may be used to collect the mapping information. Once mapped, a mock conversion may be run to determine how the data will be converted.

It is critical that your management team devote the necessary internal resources to reviewing the output of the mock conversion (data verification) to verify that existing data is being properly converted. Any instance where the data has been incorrectly converted should be investigated. It may be a simple mapping error. The issue may also be a symptom of a much larger problem: corrupt data. Address information illustrates one potential pitfall. If your current standard requires that users enter the street address on address line one and post office box information on address line two, the data will be mapped to the new solution accordingly. On existing customer records where the post office box information is placed on address line one and the street address is placed on address line two, the mock conversion will show an exception. In cases of corrupt data, the remedy is to query and scrub the existing records prior to the conversion. This process is vital to the success of the conversion and should never be deferred for any reason. Errors should be communicated to the vendor's implementation team as soon as they are discovered.

Vendor Interfaces – One of the most difficult and time-consuming tasks within the implementation process is often the coordination of other vendor interfaces. Depending on the scope of the implementation, the number of external vendors that have to modify or even build a new interface to the solution being implemented can be daunting. The internal project manager working with other members of the management team will be key to communicating, scheduling, coordinating, and implementing these interfaces. Contacting external vendors early in the evaluation process will allow you to determine the cost and timeline required by each. Sharing the primary project timeline with them once the project planning has started will ensure that they are aware of schedules and deadlines. Some interface development and implementation can occur alongside the primary solution implementation. Some will be required prior to the implementation so that they can be tested during the mock conversion process. Timelines and resource allocation planning will make certain that all peripheral solutions are functioning after the conversion date.

Halting the Implementation – If at any time during your organization's project meetings significant issues start to surface that could negatively impact the successful implementation of the new solution, your management team should consider halting the project. Unforeseen roadblocks can occur in any project. These may be related to such things as unexpected hardware delays, discovery of a corrupt database, ineffective training, or mock conversions with an abnormally high volume of data errors. The vendor will want to press on as they have other clients to implement and a master schedule to keep. As stated earlier, it is your management team's responsibility to ensure that the implementation is successful within your organization. If things are not going as planned, stop the project or verify that all parties are doing what they need to in order to bring the project back on schedule. Unless there is a critical time constraint, you are better off aborting the scheduled implementation date.

Doing the implementation right the first time is critical as there are no "do-overs." A poorly managed implementation sets a bad tone for the remainder of the time that you are on the solution. If the conversion date is critical, you should contact the vendor as soon as you become aware that your organization is no longer on a successful path. Both the solution provider and your team must be prepared to commit the additional time and resources necessary to bring the project back in line for a successful conversion.

The vendor is the primary technical resource for the solution implementation project. They should have complete knowledge of what technically needs to be accomplished in order to implement the solution successfully. They will not be able to identify all of your organization's internal requirements and interfaces that need to be included in the project. They also cannot motivate your internal staff to proactively accomplish such tasks as validation of converted data and training on the usage of the solution. You should expect the vendor to take charge of the following items unless otherwise contracted:

- Knowledge of the components and tasks necessary to implement the solution
- Configuration of existing system controls
- Conversion of any existing data to a format compatible with the new solution
- Installation of supporting hardware and software (if contracted)
- Training resources, materials, and in some cases training facilities
- Coordination of vendor implementation resources/project management

Knowledge of Components/Tasks Necessary for Conversion – As indicated previously, the vendor has the experience of past implementations. They should have project plan and statement of work (SOW) templates that can be used as a starting point for your project.

During the preproject planning process, it is critical that requirements unique to your organization's structure and operating environment be identified and included in the plan. Once the project documents are completed, they should be used as a baseline for the project. Agreed-upon changes to the project (scope changes) should be included in both the project plan and the SOW. Deviations from the project plan should be resolved as quickly as possible.

Configuration of System Controls – The vendor should be the expert concerning the setup of hardware and software controls required to optimize the operation of their solution. When possible, the primary vendor should be tasked with managing the configuration of all system controls. Doing so will narrow the accountability should the solution not operate as sold. Internal team members who will be tasked with ongoing solution operation and maintenance should be involved and understand the settings.

Conversion of Existing Data – In most cases, the vendor will use conversion programs to convert and move data from existing systems into a format applicable to the new solution. As indicated earlier in this chapter, the existing data will be captured as is, so changes must be made prior to the conversion process to avoid future problems with system operation. During the "mock conversion" process, your internal team should carefully review the resulting reports to make sure that your existing data is being properly converted. Do not assume that all implementation projects are the same. Every database has the potential of being different.

Installation of Supporting Hardware & Software – The responsibility for installation and configuration of supporting hardware and software should have been determined during the contract phase of this process. In most cases the primary vendor will install and configure the hardware to ensure that the solution will operate properly. Depending on the solution, your organization may be better off obtaining the hardware from a local vendor and having them configure the product according to the primary vendor's specifications.

In either case, the contract should clearly identify who has responsibility for installation and configuration of these systems. If this was not part of the primary contract, your management team will need to work with the vendor to ensure that the required infrastructure is in place according to the project timeline. Regardless of who is tasked with the hardware/software configuration, your management team should make sure that system configuration is on the project plan and identify who has been assigned this task.

Training Resources – Depending on the complexity of the solution being implemented, the vendor may have several different options to meet your training needs. Web-based training is a popular way to expose users to the new solution with limited effort required for setup. Some vendors provide automated online lessons and testing; these systems may include the ability to track and log training hours. While online training is a good start, it should never be relied upon as the sole training method. Ideally, classroom training should take place prior to the implementation of the new solution. This will give the trainers an opportunity to gauge the knowledge level of your team and to identify those who may be in need of additional training.

Coordination/Project Management – The vendor should assign a project manager to coordinate and manage your implementation project. This individual will be responsible for coordinating vendor resources, managing tasks according to the project timeline, and communicating the project status to your organization. The vendor's project manager will be assigned by the vendor to ensure that the project stays on schedule and the list of tasks in the SOW are accomplished. Unless otherwise contracted, the vendor will not manage the tasks required within your organization. An individual or group within your organization should be assigned to coordinate and manage the tasks that your team needs to accomplish. They will work in tandem with the vendor's project manager to ensure that all tasks are completed on schedule.

As stated at the beginning of this chapter, your management team is responsible for successfully implementing the solution within your organization. The vendor will most certainly manage the technical processes necessary to make the hardware or software run in your environment. Only your management team can identify the unique challenges and changes within your organization that must be managed in order to successfully implement a new solution.

Chapter 11

– Implementation of Customer-Facing Solutions

Critical Points:

- Implementation of a customer-facing solution requires that a plan be developed for implementing the solution internally (to your team) and externally (to your customer base).
- The impact to your customer base can be minimized through a well-thought-out rollout plan, effective communication, and proactive support.

Technology solutions that facilitate customer interaction with your organization, or "customer-facing solutions," are perhaps the most complex to implement. These include solutions such as online banking, the website, voice response (VRU), online financial management, bill payment, remote deposit capture, and mobile banking.

Customer-facing solutions actually require a dual implementation plan. As with any new technology solution, you must adequately plan the internal implementation by covering all of the components discussed in the previous chapter. Customer-facing solutions also require a plan to implement the solution externally with your existing customer base. This second level of complexity brings with it a new set of concerns. The following areas should be addressed within the implementation plan for a customer-facing solution:

- Education
- Communication and marketing
- Customer implementation process
- Support (conversion and ongoing)
- Follow-up and feedback

Education – Educating your internal team on how to properly position the new solution within your customer base is critical. Depending on the scope of the change, this could be a simple verbal message such as, "Did you know that we are now offering the ability to pay bills online?" On the other hand, this could take the form of a major train-the-trainer initiative that positions your team to demonstrate a new business financial-management solution to customers in the branch or on the road.

Prior to launching any educational initiative, ensure that your team has the information that it needs and feels comfortable enough to accomplish the goal of educating your customers. Misinformation or a poorly delivered message can be very difficult to overcome. Team members who halfheartedly endorse the new solution will pass that feeling on to your customers. Regardless of what avenue you choose for your team to educate your customers, the message must be in sync with the organization's marketing message.

Communication and Marketing – Communication with your customer base should begin well before the implementation of the solution. The communication should come in several different forms to reach as many customers as possible. Statement stuffers, messages on statements, letters, e-mails, website messaging, lobby banners, newspaper ads, billboards, radio and television ads, segments on the local television news, and verbal customer interaction have all been used by organizations in the past. Your marketing message should tell your customers what is in it for them. Sell them on how the functionality will make their lives easier, help them to better track where their money goes, or allow them 24/7 access to information. Ongoing communication of changes and enhancements resulting from technical challenges should also be planned.

Customer Implementation Process – The customer implementation process will include specifics relating to the solution rollout. If customers will need new sign-on or PIN information to access the new solution, how will that information be securely communicated to them? If the new solution requires new equipment to be delivered or software upgrades to existing equipment, how will this be handled? Individuals who will handle these tasks should be identified. Procedures should be developed to ensure that solution standards are uniformly implemented. And a timeline should be established to verify that the required tasks are completed on time.

Support (Conversion and Ongoing) – After a customer-facing solution has been implemented, the initial support can be a nightmare. Initially, those customers who failed to see any of the communications sent by your organization highlighting the change will call or come into offices to ask why you did not tell them about the change. Unforeseen technical challenges will surface and have to be resolved. Initially, some customers will be unable to use the new solution due to the "human factor". The "human factor" is the inability to adapt to change of any kind. All of these things can doom a new customer-facing solution if the support is not well planned and executed. Negative feelings created during the initial rollout can be difficult to overcome.

Follow-up and Feedback – Follow-up and feedback are essential to the ultimate success of the new solution. If the underlying business goal for the solution implementation was customer retention, do the reports of customers being added versus those being closed reflect a success or failure? Are the customers happy with the solution? What functions are they using and what could make the solution better? What percentage of the customer base is using the solution? Do we need to increase that number? If so, how? As with any technology solution, the users are the experts concerning what is right and what is wrong. Customers should constantly be surveyed to determine if customer-facing solutions are accomplishing their goals.

The challenges related to the implementation of customer-facing solutions are unique. When implementing internal technology solutions, you have a captive audience. Impacted users are internal and they should have a common desire to use the solution to accomplish the business goals of the organization. Implementing customer-facing solutions also includes a much larger external group. Your customers have high expectations, no common goal to ensure your success, and the ability to move their business to a competitor. Motivating this group to accept the new solution has many challenges. Some of these challenges are:

- Minimizing customer impact
- Resistance to change

- Downtime
- Transfer of existing customer sign-on or PIN information
- Transfer of customer account transaction history

Minimizing Customer Impact – With any change in customer-facing technology there will be impact to the customer. The degree of change will dictate the level of planning necessary to appropriately minimize this impact. A website redesign will have a much greater impact on your customer than simply replacing an existing voice response system and not changing the voice prompts. The former will require a carefully designed rollout plan; the later may simply require a notice to the customer indicating that you are updating your hardware.

Resistance to Change – If you have ever tried to change a fee, a procedure, or anything that touched your customer, you understand the issue here. Resistance to change can in some cases be overcome by showing added value, regulatory necessity, or improvement as a response to previous customer requests. Depending on the scope of the pending change, it may be advantageous to proactively contact your VIP customers to get their buy-in prior to publicly launching the project. Form a focus group and steer the results toward the features and functionality of the new solution. People tend to support what they had a hand in creating. These individuals may also become champions of the solution within your market.

Downtime – New solution implementations increase the potential for downtime. You should plan on this and determine your options for alternate processing and customer communication. Proactively addressing solution downtime minimizes the negative impact on most of your customers. They will be more forgiving if they know that you have a plan to fix the problem. This may also be an opportunity to begin the process of building a functional business recovery plan.

Transfer of Existing Customer Sign-on or PIN Information – The implementation of any new technology solution requires converting data from the existing system to the new one. When you are dealing with the customer's sign-on for a customer-facing solution, this requirement is even more critical. The inability to convert a customer's user ID, password, or PIN information from the existing solution to the new solution means that the customer must be involved in the conversion process. Customers will be required to use a temporary sign-on or PIN to access the new solution and re-establish their original sign-on or PIN. On the surface this may not sound like a significant issue—but it will be. Every effort should be made to convert the customer sign-on or PIN as it exists on the current solution.

Transfer of Customer Account Transaction History – Account transaction history is less critical than converting the sign-on or PIN information—but it is not too far behind. In most cases the current solution will no longer continue to function once the new solution has been implemented. If account transaction history cannot be migrated to the new system, your organization will need to come up with a way for your customers to access this information. The more automated the solution, the more difficult it will be to resolve this issue. As with the sign-on information, every effort should be made to convert the account transaction history to the new solution.

Chapter 12

– Supporting Another Solution

Critical Points:

- An internal resource should be assigned to manage the solution and all of the tasks required to ensure its success within your organization.
- Once you have determined what impact the solution has on your operations, your management team should review and revise internal processes and procedures.

After the dust has settled from the implementation, your organization is going to be left to support yet another technology solution. If you identified a member of your team to shadow the implementation process as recommended in the previous chapter, you are ready for the challenges that this new solution will present. This individual will act as an internal sponsor for the solution, interacting with users and the vendor representative and facilitating the tasks necessary to successfully manage the new solution.

There are five keys to successfully supporting the new solution:

- Knowledge of the solution
- Customization of processes and procedures
- Communication to/from users (also customers if customer facing)
- Issue resolution
- Vendor relationship

Knowledge of the Solution – If the individual assigned to support the new solution does not have a working knowledge of the product, he or she will not be successful in supporting the solution or its users. Even if your organization is small, you must find the human resources to staff this position. You may have asked the vendor if they provided daily support of the product during the evaluation process. The sales professional may have emphatically assured you that they did; however, your vendor is probably not staffed to support your team with the types of questions and issues that come up with daily usage. If your expectation is that the vendor will fill this role, you are going to be disappointed. The resulting user dissatisfaction will negatively impact your organization's ability to optimize your investment in this solution. You should have an individual or a group of individuals within your organization who can handle the more common daily user questions and issues. One person from within this group should then be responsible for referring more complicated questions and issues to the vendor. Think of this as a tiered support structure where the first level of support is internal and the second level of support is with the vendor.

Customization of Processes and Procedures – Once you have determined what impact the solution has on your operations, your management team should review and revise internal processes and procedures. Some of this information may be known prior to the implementation through your communications with the vendor. If you have not already done so, contact other power users of the solution and ask them to share their documentation. If they are willing to do so, you will simply have to merge your processes and procedures with theirs. If they are not willing to share, or they have nothing in place, ask if you can send someone to their location to see firsthand how they are using the solution. Then offer to share your finished product.

Communication to/from Users – The third key to successfully supporting a new solution is communication. The individual within your team who is supporting the new technology must be able to effectively communicate with the users of the solution. This includes communicating initial and ongoing training, vendor fixes, vendor-generated enhancements, and management expectations of the solution's use.

One of the ways to accomplish this in larger organizations is to facilitate periodic regular meetings. In them, representatives from each location should get together to discuss issues related to the usage of the solution as well as improvements that have been made by the vendor. Think of it as a mini internal user conference. Participants should be encouraged to come to the meeting prepared to talk about issues related to solution usage as well as the positive benefits. Too often users focus on the negative when they talk about a solution. This forum should be about both. A positively focused agenda could include the following points:

- Introductions of the team members (if necessary)
- Restatement of the business goals for the solution being discussed

- Evaluation of the achievement of business goals through solution metrics
- Discussion of improvements that have been made by the vendor
- Discussion of current issues and an action plan for reporting to vendor
- Discussion of positive benefits of the solution
- Discussion of negative impact of the solution along with a mitigation plan
- Adjournment

When the solution is customer facing, a process for customer communication must be developed and maintained for the life of the solution. Communication in advance of pending hardware or software updates requiring solution downtime will prevent customers from developing a negative opinion of the solution. Customer feedback is also necessary in order to ensure that the solution stays relevant within your market.

Issue Resolution – An individual within your organization must be identified to the vendor as the go-to person when it comes to solution support. This internal resource will be the primary conduit for the reporting of issues to the vendor. This way, your management team can monitor the level of service and the volume of issues. This is especially important when the vendor contract has service level agreement benchmarks. It also keeps the same issue from being reported by multiple internal users. While this may seem like the vendor's problem, it actually clogs the support system, causing delays in addressing your more serious issues. By routing all issues thought this one internal support individual, he or she will continue to expand their knowledge of the solution. In time, he or she will be better equipped to provide more in-depth internal support, speeding the process of issue resolution. This individual should also be attending vendor user conferences to build contacts and to make the vendor aware of user feedback.

In addition to reporting issues, the organization should maintain a record of issues that have been reported along with the resolution status and other historical information. Some vendors have an automated process in place for you to monitor this information. If not, a simple spreadsheet can be maintained with the following information:

- Date issue reported
- Issue description
- Relative severity rating
- User who reported
- Current resolution status
- Resolution description

Keeping track of solution-related issues and their resolution will allow your management team to monitor how effective the vendor is at resolving user issues. It also allows for an assessment of the possible productivity costs associated with the issues being reported. Users who frequently report support issues may be in need of training; they may also be identifying a significant problem with your organization's usage of the solution.

Vendor Relationship – The relationship that your organization has with the vendor is crucial to facilitating the flow of feedback to your vendor and the flow of solution-related information to your organization. We will talk more about this in Chapter 15.

Chapter 13

– Follow-Up and Return Analysis

Critical Points:

- After the new solution has been implemented, the organization should conduct a follow-up analysis to ensure that the solution is being used efficiently and that the new technology is indeed helping accomplish the organization's business goals.

- If follow-up review reveals that the solution is not facilitating a positive impact to the organization's business goals, an effort should be made to determine and correct the root cause.

Too often, an organization goes through the effort to implement a technology solution just to find out later that it was not as successful as envisioned. Even when great care was taken to select the ideal solution, the focus is lost once the solution is installed. The organization must monitor the ongoing success of the solutions that it implements. For example, if a retail delivery solution is not entirely successful, it could be due to inadequate advertising, poor employee training, poor vendor support, or technical issues related to its operation. Whatever the case, continued focus would have allowed the management team to make adjustments to achieve the expected return and to maximize the benefits of the new solution.

After the new solution has been implemented, the organization should conduct a follow-up analysis to ensure that it is being used efficiently and that the new technology is indeed helping accomplish your business goals. One of the best ways to do this is to have the vendor conduct a post-implementation review. In many cases, this will require that the vendor send a representative on-site to talk to your team and to observe their workflow. While this process may cause the organization to incur additional travel cost, it is a worthwhile exercise. It will help ensure that the users are adhering to workflows and operational procedures that will ensure success with the investment made in the solution. The alternative is to wait until issues are apparent, in which case it may be too late to break old habits. All too often this step is delayed or skipped altogether, resulting in both short- and long-term negative implications.

You never know what you don't know until you begin using the solution. For this reason, an additional option for establishing optimal usage is to visit another vendor client using the solution in an operational environment similar to yours. The vendor should have given you information pertaining to these clients during the reference phase of the selection process. It is best to send individuals with a good knowledge of current operational processes and procedures within your organization; still, they should have an open attitude concerning change and new procedures. Their mission is to review operational workflow at the seasoned user against current practices at your organization. This will ensure that the proper controls and procedures are in place to make certain that your users will succeed.

The second part of the post-implementation review is to measure the return on investment (ROI) against the business goals used to justify the solution purchase. As indicated earlier, some examples of return are:

- Increased revenue
- Improved customer service
- Enhanced retail delivery
- Increased efficiency
- Account growth
- Account retention (market pressure)
- Regulatory compliance

With a few exceptions, such as growth or regulatory compliance, the return may not be immediately noticeable, or it may be difficult to measure. This evaluation, however, is very important. Some results may take a year or more to become apparent. This lead time should have been estimated within the business case along with accountability for the follow-up and review. Results should be reported back to the management team at regular intervals.

If the follow-up review reveals that the solution is not supporting the organization's business goals, an effort should be made to determine and correct the root cause. Possible issues could include:

- A misalignment of the solution to the business need
- Improper configuration during the implementation
- Inadequate initial or ongoing training of users

- Operational processes that do not optimize solution usage
- Inadequate user support
- Subsequent changes to market or operational workflow
- Changes in the regulatory environment

A Misalignment of the Solution to the Business Need – A misalignment of the solution to the business need is rare if the methodology discussed previously is used to evaluate and select the solution. If it does happen, the business case review process should be evaluated to determine where the breakdown in the benefit analysis occurred. If the product was properly implemented and is being used as intended, there should be some positive, measureable impact to the organization. If there is not, a solution re-evaluation should be considered.

Improper Configuration During the Implementation – Improperly configured solutions can be the result of several causes. In many cases it is difficult for your management team to know with certainty what configuration will optimize usage of the hardware or software until after the solution has been implemented and is being used. The vendor can make suggestions; however, they do not know your environment. If adjustments were not identified and made after conversion, the issue could negatively impact the solution output.

Miscommunication between your organization and the vendor could also be to blame. In either case, if the error is not caught during the post-conversion review, it can negatively impact the solution output. The vendor should be contacted to determine if the solution configuration can be altered. If this is possible, your management team should review the impact of all configuration changes prior to implementing them. Unfortunately, even minor changes made after a solution is operational may result in unintended consequences in other areas of the organization. If additional configuration changes are not an option, there may be procedural changes that can be made to optimize the solution's impact.

Inadequate Initial or Ongoing Training of Users – Inadequate user training is one of the most common reasons for a solution not being as effective as it could be. Not sufficiently training users during the implementation process, or not reinforcing that training periodically after the implementation, will doom even the best of solutions to poor performance. Your internal solution sponsor should be responsible for conducting periodic meetings with users representing all of your locations. High-level (train the trainer) topics should be covered during these meetings with the intent that the information be relayed to all corresponding users by the location representatives. If meeting discussions determine that additional user training is required, internal or external (vendor) resources should be procured and an organized training initiative should be developed and implemented.

Operational Processes That Do Not Optimize Usage – When the organization does not take the time to evaluate operational processes and procedures during the implementation phase of the project, the workflow will not be adjusted to optimize usage of the solution. Since the daily procedures no longer match the flow of the new solution, team members are prone to looking for "work-arounds" or to "fooling the system" to solve what they perceive to be system deficiencies. To resolve this issue the internal solution sponsor should use the periodic user meetings to facilitate an ongoing review of processes and procedures. The users will know better than anyone what tasks within the daily workflow conflict with system operations.

Inadequate User Support – Inadequate user support can result when there are no internal controls to monitor resolution of user questions and concerns. Without a way to quantify the level of user support being provided internally and by the vendor, your management team has no way to manage this critical process. Poor user support can be remedied by having the internal solution sponsor and the primary vendor representative get together to discuss more effective ways to address user issues. A process should then be put into place to monitor the ongoing resolution of user issues with feedback being given monthly to the management team.

Subsequent Changes to Market or Operational Work Flow – Changes within your marketplace subsequent to the implementation of the solution, or changes to internal workflows resulting from other projects, can alter previously optimized processes and procedures. Discussions during periodic user meetings should help to identify workflow changes that conflict with solution usage. When these are identified the management team should work to revise the operational practices once again.

Changes in the Regulatory Environment – Changes in the regulatory environment can minimize the positive impact a solution has on your business goals. Fortunately your organization is not alone in this scenario. Other organizations are also faced with the issue of maximizing their technology investments in a regulatory environment that appears to be constantly changing. For this reason you can be sure that there are many heads working to optimize a bad situation. Your vendor or other solution users should be contacted to determine what actions they are considering to mitigate any negative regulatory impact. Regular attendance of vendor user conferences will also allow your management team to be better prepared to proactively deal with these challenges as they occur.

Chapter 14

– The Solution Life Cycle

Critical Points:

- The decision to stay with an existing vendor or to search for alternative solutions should be part of an annual review based on your strategic technology plan.
- Your market demographics, strategic business goals, business evolution, and current technology determine the length of the technology solution life cycle.

All too often we have a technology solution that has faithfully served us for many years, and even though the system is no longer as effective as it used to be, we can't imagine it not being there. Even worse, we cannot imagine having to make a change. The reality is that technology solutions have a life cycle. At some point every solution comes to the end of its useful life, or stops being as effective as it could be. When that happens, your organization must decide whether to stay with a current solution or look for a new one.

The decision should be part of an annual review based on your strategic technology plan. This decision should also be based on projected business needs and not just timing of contract renewal. Making the decision to change vendors well in advance of the existing contract renewal date will allow your management team the time required to evaluate other vendor solutions and to make an informed cost-versus-benefit decision.

The technology solution life cycle is determined by your market demographics, business evolution, strategic business goals, and the type of technology that you are using. Over the history of technology development, the life cycles of solutions have grown shorter. Products such as ATMs are now in the maintenance phase of their life cycles with some vendors working on additional development in an effort to breathe new life into the solution. The life cycle of the ATM has spanned a little over forty years with very slow adoption and maturity phases. Other products, such as mobile banking, are still in the early adoption phase. The mobile banking life cycle is as yet unknown, but the adoption and maturity phases of this product are projected to be much shorter than that of previous technologies. Your management team should factor this into your strategic technology plan and into your decisions to invest in such solutions. Once you elect to adopt a new technology, be prepared to budget the cost of staying current. Buying a newer technology and then falling behind is not an option if you wish to maximize your investment. The typical technology solution life cycle has the following four phases:

- Introduction and early adoption
- Mainstream adoption
- Additional development
- Maintenance

Introduction and Early Adoption – The introduction and early adoption phase is usually characterized by high costs, little competition, low customer demand, and minimal return on your investment. Potential customers must be incented to try the product, while marketing costs related to product introduction are high. Commonly referred to as the "bleeding edge," your organization can expect frequent product updates and refinements by the vendor as they correct unforeseen development issues. There is a high degree of reputation risk as some negative impact to your customer base is possible. Vendors are usually willing to make deep discounts on solution pricing and may be willing to work with your organization to offset some of your costs as they push to get early adopters in the market.

Mainstream Adoption – As mainstream adoption takes over, customer awareness has increased and the solution becomes a requirement within the marketplace. Marketing costs and user incentives are still a factor but they are less important. The solution receives frequent attention in the press and at industry trade shows. Vendor prices are typically high as the demand for the solution has increased. Solution problems are minimal as the early adopters have worked out the most common bugs.

Additional Development – As the solution continues to mature, the vendor undertakes additional development to extend the useful life of the solution. Regulatory and marketplace changes are made; suggested enhancements by current users may also be incorporated. The cost of the solution purchase and introduction may moderate during this phase as the market becomes saturated. Implementing the solution at this point in its life cycle will have minimal impact on the ability to gain market share. If demand continues to be constant, the solution may, however, facilitate additional revenue generation.

Maintenance – The maintenance phase usually marks the point in the solution life cycle where technology, evolution, and the marketplace have moved on. The vendor continues to maintain the solution for those who choose not to evolve. Continued development and enhancements are usually relegated to regulatory and security corrections.

Knowledge of the projected customer demands of your marketplace, as well as an awareness of the latest technology developments, is essential to managing the life cycles of your current technology solutions. Where are your current solutions in their life cycle? What solutions do you need to be watching? When should your organization adopt? All of these questions should be asked often as you constantly evaluate your strategic technology plan. Answers to these questions can be found by reading industry publications, attending conferences, interacting with other organizations within your industry, or through the services of a qualified technology consulting professional.

Chapter 15

– The Vendor Relationship

Critical Points:

- You should only consider doing business with vendors that have a positive reputation for being good business partners and not just vendors.
- Develop a close working relationship with the vendor and participate in their user conferences and development planning.

Regulators are increasing the focus of their technology exams on vendor management. More specifically, you must have a plan for understanding, measuring, monitoring, and controlling risk as it relates to your key vendors. You must also have knowledge of their development road map for the solution that you have purchased. Even though much of this information can be obtained directly from the vendor, the more critical knowledge comes from having a personal relationship with the vendor representatives. Developing and maintaining this type of relationship will help reveal the good and the bad about your vendor and their solution.

A good partnership with your technology vendor is also critical to maximizing your investment in their solution. Adversarial client/vendor relationships benefit neither party and will almost always lead to poor usage of the technology solution. As noted in the chapter on vendor selection, you should only consider doing business with vendors that have a positive reputation for being good business partners and not just vendors. Of course, the responsibility for having a good client-vendor relationship resides with both parties.

One of the best ways to promote a healthy vendor relationship is for your management team to attend user conferences and forums sponsored by the vendor. While there is always an expense related to these types of activities, that expense should be considered as a cost of the solution and should be budgeted annually. Once your management team makes the decision to purchase a solution, you owe it to your organization to maximize the benefit of your investment. The only way to accomplish this is to be an active user and to continue your investment in the solution. Conferences and forums allow your organization the opportunity to interact with vendor management, knowledgeable vendor representatives, and other vendor clients. Frequently, user-to-user exchanges allow you to identify uses of the software or hardware that even the vendor may not know about. Policies and procedures may also be exchanged, saving your organization from having to repeat the process already undertaken by previous users. The registration fees and travel expenses can be more than offset by the benefits of attending. One caution here: Attendees must actively network and participate. The vendor only sponsors the event. What your organization gets out of it is solely up to the participants. If you have an individual who returns from a vendor conference saying that it was not worth his or her time, do not abandon the event. Send someone else to the next conference and react based on the additional feedback.

Your management team should also have a good personal relationship with an individual representing the vendor. Most vendors assign a Relationship Manager or Strategic Account Manager to each client. It is vital that your management team gets to know this individual personally and work with him or her to maximize your usage of their organization's resources. As with any human relationship, there can be conflicting personalities. If this is the case, request a change in relationship managers and find someone with whom your organization can work. The more closely you can work with this individual, the better your experience will be with both the vendor and their solution. We all tend to do special favors for those we know well. This relationship is no different. If your organization is large, appoint one member of the senior management team to "sponsor" the vendor and act as a liaison with the relationship manager. This person must understand the importance of his or her role in managing and cultivating the vendor relationship. While it is a business relationship, do not underestimate the value of the personal aspect in the partnership.

Once you have developed a close working relationship with the vendor, ask to be part of their solution planning and development process. By doing this you signal that your organization intends to be a part of the vendor's future and that you are willing to contribute to their strategic direction. Your actions will encourage the vendor to take special notice of you as a valued client. In turn, the vendor will want to invest in their relationship with your organization. Think about it: Would it not be easier for your organization to plan the next few years if you knew what your customers were thinking now? What would your reaction be to customers who actively contacted you and said that they wanted to help you to understand their future needs? This process will strengthen your relationship with your vendor and it will go a long way toward ensuring that they are creating solutions that meet your needs.

You may be saying to yourself, why should I spend so much time trying to form a good personal relationship with the vendor? Should the vendor not cater to me? After all, I am the customer. Fortunately that is the thinking of most of the vendor's clients. By carefully choosing only vendors with a reputation for being good technology partners, and reaching out to them to proactively form a great partnership, these vendors will come to value you as a client above others; they will have an increased interest in keeping you happy. As with any kind of relationship, you will get out of this one proportionate to what you give to it.

Chapter 16

– Who Has Time for All of This?

Critical Points:

- Technology consultants bring a specialized skill set and knowledge base to your organization that is focused on managing your technology investments to meet your business needs.

- The consultant's knowledge of technology industry standards, vendor management practices, technology project management, and market focus can help to guide your organization's strategic technology decisions.

Marginally successful solutions are the result of an organization not investing the time and effort required to properly select, implement, and manage solutions that add to your organization's business goals. The planning and management processes discussed here are necessary due to the speed at which technology changes. Missteps and mistakes can cost your organization dearly when you are locked into a long-term contract for a technology that becomes obsolete or fails to meet your needs. To compete in the technology arena you must keep your eyes on the horizon and monitor the success of the next "killer app." You must constantly evaluate the effectiveness and risks associated with your current solutions and ensure that these things are being properly managed. You should also maintain a close working relationship with your current vendors to ensure that they are moving in a direction that will keep your organization in the game. When a replacement solution or a new solution needs to be purchased, it is up to your management team to effectively research what you are buying and to fully implement it within your organization. While these tasks are not difficult, they are time consuming. When broken down into a list of individual tasks, the process looks something like this:

- Create and maintain the strategic technology plan
- Monitor risks associated with your technology solutions
- Provide user support of current solutions
- Monitor solution lifecycles
- Attend user groups and monitor technology developments
- Develop and maintain working and personal relationships with vendor representatives
- Facilitate annual review of vendor's performance
- Sponsor internal user group meetings and training opportunities
- Facilitate ongoing ROI measurement and evaluation
- Facilitate the annual technology budgeting process
- Facilitate the business case review process (when necessary)

- Facilitate contract negotiations (when necessary)
- Create RFPs (when necessary)
- Manage implementations (when necessary)

If your organization is large, you probably have a senior-level manager who will be responsible for ensuring that these tasks are accomplished. In smaller organizations, the tasks may be assigned to a midlevel manager as an additional set of duties. In either case, your organization may want to consider using an outside technology consultant to handle the strategic and technology management tasks. Having a technology professional on your management team will allow you to have the same edge as your larger competitors. You will be kept up to date with the latest products, knowing when to purchase and when to hold back. A strategic technology consultant provides smaller organizations with Chief Information Officer (CIO) level services that larger companies enjoy, but at a reduced cost. Using a technology consultant provides:

- A dedicated technology resource
- Facilitation of discussions by an impartial third party without internal alliances
- Knowledge of alternate best practices being used by other organizations
- Coordination of technology projects
- Knowledge of the industry and vendors
- A way to minimize the impact of technology management on organizational resources

- Independent senior-level accountability
- Knowledge of current vendor pricing structures
- Risk management expertise
- Enhanced credibility to examiners

A technology consultant should be considered a trusted advisor to your senior management team. He or she brings a specialized skill set and knowledge base to your organization that is focused on managing your technology investments to meet your business needs. The consultant's knowledge of technology industry standards, vendor management practices, technology project management, and market focus can help to guide your organization's strategic technology decisions. Due to the level of confidence and trust that is required of this type of relationship, it is critical that the consultant interact well with all levels of your organization, especially the senior management team.

Technology consulting firms offer a variety of services that address solution relevance. Smaller consulting firms specialize, and are appropriate when your needs are focused. Larger firms offer a broad assortment of services such as strategic technology planning, external IT exams, penetration testing, network design, risk assessments, Web design, and business continuity. These firms are more appropriate for larger organizations with broader needs and larger budgets. You should choose a consultant who can actively contribute to your organization's management team and who understands your strategic business planning process. In order to maintain an ongoing knowledge of organizational direction and to provide input, the consultant should be included as a member of your management team attending strategic and operational meetings.

At the end of the day, the responsibility for having the right technology solution rests on the shoulders of your organization's management team. With the help of the concepts within this book, it is time to start controlling your destiny when it comes to technology.

Appendix A

– Strategic Technology Plan Outline

Conduct an annual evaluation of current technology vendors that encompasses:

I. A process to monitor key contract dates

   a) Non-renewal notification dates

   b) Renewal date

II. A comprehensive GLBA risk analysis

   a) Strategic risk

   b) Operational risk

   c) Reputation risk

   d) Compliance risk

   e) Financial (credit) risk

III. A review of performance against business goals (ROI)

a) Increased revenue

b) Improved customer service

c) Enhanced retail delivery

d) Increased efficiency

e) Account growth

f) Account retention (market pressure)

g) Regulatory compliance

A plan for ongoing support of existing solutions:

I. User support

a) Enhancement development

b) Regulatory compliance

c) Service level agreements (SLAs)

II. Implementation and dissemination of enhancements

a) Hardware enhancement implementation plan

b) Software enhancement implementation plan

c) Hardware system updates

d) Software release update

III. A service level evaluation

a) External user support

b) Internal user support

IV. Required hardware or software changes

a) Sunsetting of solutions

b) Technology evolution

A plan for the evaluation of future technology purchases:

I. Business case discipline

a) Business need

b) Anticipated outcome

c) Impact to organization

d) Risk analysis

e) Expected return

f) Projected cost

II. Technology budgeting process

a) Justification for ongoing technology expenses

b) Approval of additional technology projects

III. Compatibility and support of strategic business objectives

a) Awareness of technology and market trends

b) Review of regulatory requirements

Appendix B

– Business Case Worksheet

Date: ______________________

Solution: ____

Business Case Sponsor: ______________________________________

Solution Description
Description should be detailed enough for the entire management team to understand what solution is being considered and its impact to the organization.

Business Need
How does this solution contribute to the organization's business strategy? What business needs will this solution address?

Estimated Impact to Current Operations
Describe the expected impact that the solution will have on current operational processes, workflows, controls, training, compliance, and risk mitigations.

Projected Return on Investment
Describe the expected measureable return to the organization once this solution is implemented. Also, indicate the projected lead time before the return should be measured.

Estimated Time and Resources for Implementation
Estimate the individuals who will be responsible for the implementation of this solution along with the projected time that it will take them to complete their assigned duties.
Preproject Planning: System Configuration – Planning/ Review: Training of Users: Installation of Hardware/Software: Testing of Conversion Files: Conversion of Existing Data/Review: Interface Testing: Solution Rollout/Promotion: Marketing: Project Wrap-Up:

Hardware/Software Requirements
List hardware and/or software required to operate the solution. Include estimated cost and the lead time required to install the hardware/software.

Key Resources Required
Include the names and estimated involvement of key individuals within the organization who are critical to the success of the project.

Estimated Total Cost
Include estimated costs relating to purchase of the solution, implementation, support, conversion, deconversion, training, communications infrastructure, interfaces, marketing, and promotion.

Assessment of Impact to Organizational Risk
Projected impacts to strategic, operational, reputation, compliance, and financial risk should be estimated here.

Business Case Review Date: ________________________

Business Case Review Committee Conclusions/Recommendations:

____

____

____

____

____

____

Additional Information Required:

____

____

____

____

____

____

Next Steps:

____

____

____

____

____

____

Business Case Approved By:

____

Date Approved: ______________________________

Reason for Denial (if rejected):

____

____

Appendix C

– Request for Proposal Outline

I. Statement of purpose

a. Description of need

b. Objective of solution

c. Business goals being addressed

II. Background information

a. Mission statement

b. Asset size

c. Number of users

d. Market demographics

III. Operational requirements

a. Workflows

b. Transaction types

c. Transaction volumes

d. Operational handling requirements

e. User interfaces

f. Balancing requirements
g. Regulatory requirements

IV. Technical requirements

a. Hardware
b. Software
c. Operating system
d. Network
e. Communications infrastructure

V. Support requirements (hours/time zones)

a. Organizational user support hours
b. Customer support hours
c. 24/7 support
d. Holiday coverage
e. Hardware servicing

VI. Performance standards

a. System uptime
b. Transaction processing time frame
c. Projected growth requirements

VII. Security requirements

a. Physical
b. Online

c. Data storage
d. Encryption

VIII. Interface requirements

a. List of vendors currently being used and interface type
b. List of any prospective vendors currently under consideration
c. Any other interface requirements

IX. Business recovery

a. In-house recovery options
b. Outsourced vendor plans are tested at least annually
c. Documented/tested plan for solution

X. Required deliverables

a. Desired vendor deliverables
b. Timeline requirements

XI. Special contractual requirements

a. Contract term limitation
b. Desired financing arrangements
c. Conditional or contingent acceptance

XII. Pricing structure

a. Software licensing
b. Hardware (lease/purchase)

c. Required operating system
d. Required communications/data transmission infrastructure
e. Installation/implementation of solution
f. Training
g. Project management
h. Post-conversion support
i. Ongoing solution support
j. Required third-party interface
k. Ancillary vendor add-ons
l. Outsourcing option
m. Maintenance and support

XIII. Vendor qualifications and references

a. Vendor history
b. Corporate culture
c. Mission statement
d. Strategic development goals
e. List of references
f. Current financial statement
g. Pending litigation
h. Number of clients currently using solution
i. Age of solution

XIV. Financial analysis

a. Audited financial statements
b. Balance sheet and income statement
c. Peer analysis
d. Industry publications/news
e. Trend analysis

XV. Administrative expectations

a. Where/to whom response should be sent
b. Timeline for submission of response
c. Format requirements
d. Next steps

XVI. Organizational contact information

a. Primary RFP contact
b. Content contacts

Appendix D

– Suggested Vendor Reference Questions

How long have you been a client of the vendor?

Overall, how would you rate the relationship?

What do you like most about the vendor?

What do you like least about the vendor?

If you had to choose a vendor again, would it be this one?

What other vendors did you consider during your selection process?

How did your conversion go? How was the vendor's conversion support?

Have you had any hardware or software issues that we should be aware of?

Are the vendor's support contacts knowledgeable?

Are the support contacts able to resolve most issues quickly?

How often do the vendor's representatives contact you proactively?

Is the vendor responsive to your enhancement request?

Are you aware of any compliance issues related to the solution?

Are you aware of any security issues related to the solution?

Have you seen any disturbing trends that we should be aware of?

How well does the solution interface to other systems?

What other systems are you currently running that interface?

Have you completed a financial analysis of the vendor and what is your impression?

CPSIA information can be obtained at www.ICGtesting.com
Printed in the USA
245194LV00001B/1/P

9 781432 763879